BORN OF BURNING EMBERS

BORN OF BURNING EMBERS

G. A. JØHN

For my mother, who would stop at nothing
to care for and protect her children.

For my family, for helping me soar towards my
dreams and encouraging me to reach for the stars.

Finally, for God, as this would never be possible
without you.

*'My God hath sent his angel, and hath shut the
lions' mouths, that they have not hurt me:
forasmuch as before him innocency was
found in me: and also before thee,
O king, have I done no hurt.'* — Daniel 6:22

Prologue

The Culling

IT ALL STARTED WITH THE BLACK RAIN, but that wasn't what wiped out the men. Second came the plague, which spread like wildfire, killing off most of the world's life. People didn't know they were dying until it was too late. The world's population plummeted, entire civilisations were erased, and ecosystems were destroyed. But that wasn't what wiped out the men either. It was the war, 'the Culling', as they called it. That's what took the men, all of the men.

The women of Earth blamed man for the prior events and declared war against each other. The Culling was only the beginning of their problems. It stole the lives of billions – men, women, and children. But near the end of the extinction, the

Culling stopped taking the lives of women; by the orders of the war leaders, only the men were hunted. Eventually, the Culling decided to spare the lives of the few thousand remaining women. It was cunning; somehow it knew that without the male sex, the human race could never sustain itself. It knew that the women would slowly die out, alone, and that the human race would ultimately come to an end.

The remaining women searched and searched for many different ways to repopulate the earth, but their efforts were fruitless. It wasn't until decades later that some women were exploring the ruins near the border between Canada and the United States, and discovered an abandoned laboratory full of pre-war research. The forgotten research explained a method of reproduction using bone marrow as the primary ingredient. Although it was arduous to replicate, the research and tests provided accurate results. For the first time after the extinction of men, there was a chance, a glimmer of hope. But it was short-lived. It was discovered that a child born of bone and cell would always be female. No male child could be created without existing male DNA. At that very moment, the idea of humankind changed radically; the male species was lost forever. For the next two centuries, man was just a myth.

During this time, women set up a new governmental structure, appointing the Weiluk family as the monarchs. They created a sustainable society without men, but some of them knew that eventually man would have to return. Nature would reset itself, and just as it took away the male species, it would return them to the world. Over the years, some prophesied the return of man, but they were shut down by their superiors. Men were taken for a reason. This was what the surviving women believed. It had become an absolute truth, and the women began to live in continuous fear should man ever return.

PART I

CHAPTER

ONE

THE PROCEDURE

J ASIRA TUCKED AWAY AT THE SIDES of her plastic
bedsheets, which crinkled and tore at the corners, and tried
to unstick them from her skin. She had made pillows from
stacks of small crates and insulation from the walls. Of course,
she had covered these 'pillows' with the same plastic as her
bedsheets to avoid irritating her face. With a quick puff, she
blew out what remained of the candle beside her bed – the wax
had spread out into a sticky goo as it melted in the previous
night's heat.

Jasira and her non-biological sister Vaika lived in The
Border, a city of ruins founded after the extinction of men.

Since the Culling, most remaining women lived in The Border, but some of the founders left to create another city, Monday, so-called because that was the day The Border split. The two cities eventually merged into one nation, Diektra, which is now ruled by the monarch. None of the royal family ever visited either of the two cities; they lived in a citadel too far away for the journey. Strangely, nobody knew where the citadel actually was. For Jasira and Vaika, The Border was home.

Jasira's birthday fell in the middle of June, when the summer winds were warm. This birthday was different, and not just because the season felt different this time round. It was her twenty-first birthday; the birthday she and every young woman in The Border dreaded. It was the day of 'the Procedure' – they called it that because nobody could remember the full name. When a girl reached the age of twenty-one, she had to prepare herself for the Procedure – childbirth – to repopulate the world. Some may call this a twenty-first birthday present, but the pain a woman went through just to fulfil the hierarchy's orders was far from it.

Jasira ambled into Vaika's room, which was much smaller than hers. She arched her back, stretching and unlocking the discs in her spine and accidentally knocking over a small wooden lion from Vaika's bedside table, if it could even be called that. The lion broke into two pieces as it landed on the carpet. She froze in an awkward position and hoped the sound didn't wake up Vaika. She reached for the pieces and tried to force them together. The lion had been a gift for Vaika's child. She would have been five years old in a few months, but something went wrong near the end of Vaika's pregnancy, woefully ending the child's life. Vaika begged for another chance, but the law only allowed one child per woman. For the law is written:

At the age of twenty-one,
a young woman must bear one child,
and one child only,
to sustain the human race until the end of time.

After decades of uncontrolled population growth, the superiors feared they would be overthrown. The law was set so that no woman would be filled with the very things that ended man – greed, power and wickedness. After the extinction, the founders believed that the Culling had let the women survive because men were unfit for positions of power. They believed that the world had given the human race a second chance – a chance to create a better world, a world without man.

Vaika rolled over on her spread of thin cardboard pieces and squinted her eyes to adjust to the early morning sunlight.

"Happy birthday, Sira!" Vaika called Jasira that for short.

"I don't think happy is the right word," Jasira responded sarcastically, placing the wooden lion back on the bedside table.

Vaika outstretched her arms and leaned against the mouldy wall. The mould had spread over time, much faster recently.

"Sira," she began, "I felt the exact same way five years ago, and if given the chance, you know how much I would give to do the Procedure again." She opened the bags she used as pillows and pulled out a small book. "This was our mother's, now it's my gift to you, a birthday present," she said, handing something to Jasira.

"A Bible?" she asked, leafing through the pages.

"It will teach you important life lessons you know, as it did for me."

"Thanks," Jasira said as she placed it on the bedside table and put on her scuffed black boots, leaving the laces untied.

Vaika brushed her teeth in their dilapidated bathroom. The basin still ran water but its purity was questionable, and the pipes never brought fresh water. Vaika, and most of the people in the city, fetched drinking water from the mountains close by.

"I wasn't as nervous as you for my Procedure," she said, splashing the water across her face.

"Is it that bad?" Jasira asked. "I'm shitting myself."

"It goes fast. You've got nothing to worry about, Sira."

"I'm just really nervous."

"Sira," Vaika said comfortingly as she tied up her long blonde hair into a low ponytail and put her leather jacket around her shoulders, "I'll come with you. I have to see the healer while I'm there anyway."

"What do you need the healer for?" Jasira asked.

"Nothing important. Look, I'll be with you the whole time." Vaika changed the topic. "Quick, get ready, fix your hair. You look like you just woke up."

"That's because I just did," Jasira mocked.

Jasira let her hair down and brushed out the knots. It was a darker brunette and much thicker than her sister's. She rushed to brush her teeth after she pulled the laces on her boots tight. She stared at her electric blue eyes in the cracked mirror, took in a deep breath, and slowly exhaled. Vaika came up from behind her and noticed Jasira rubbing the scar on her palm.

"I remember when you got that, we were so young back then," she said. "You'd scratched your hand on a sharp metal sheet on the wall. I bandaged it up real good."

"I don't remember that part. It's what happened after that I remember so clearly," Jasira began, nervously laughing. "Nobody believed me, except you."

"Nobody believed you because lions never had lived in this part of the world. But I believed you because I saw it with you in the empty street just before it ran away."

"It still confuses me how a lion cub ended up in this part of the world. They're supposed to be all the way over in Africa," Jasira said. "But I guess after the Culling, the world rearranged itself."

"The world needed rearranging."

Vaika climbed through an aluminium tunnel just big enough for her slim body to fit through. She held Jasira's wrist tightly and pulled her through the same tunnel, squeezing through the obstacles to get out. It took a moment for Jasira's eyes to adjust to the brighter light. Their home was tucked away between two towers that had collapsed onto each other, near the outskirts of the city. The remains of the potholed road were cracked, like veins slowly breaking through the concrete, and gradually worsening. Jasira's feet always got caught in the potholes; she never paid attention to anything past her nose.

"Morning, Prianaj!" Vaika waved at the woman walking her food cart past them.

Prianaj lived a bit further out of the city than them, quite a bit further out actually. She lived close to the mountains. She started growing her own fruit and vegetables after people became sick of the food they normally ate, physically sick. The royal officials ate the good stuff. Some even ate almost as much as three dogs a day, and over the years, people started calling them 'Cerberi'. The rest of the citizens only ate what they could trade at the market or they had to grow their own produce.

The others in the city tried to avoid Prianaj; she was the crazy lady nobody wanted to talk to. Prianaj separated her frizzled hair with her wrinkly fingers. Her mouth curved at the edges, and her eyes squinted slightly. Jasira stepped behind Vaika, shielding herself from Prianaj's smile.

"A special day today, isn't it?" Prianaj began as she pulled her cart over a pothole that caught its wheel. She spoke in a deep voice with a rich tone, always clear and crisp.

"Indeed, Prianaj," Vaika said. She was always the one to talk to people. Jasira never liked the attention.

"All the best, you two," Prianaj said calmly as she regained control of her cart.

Jasira waited for Prianaj to move completely out of sight before continuing. They rounded the corner of the collapsed buildings and found themselves in the main centre walkway. It was the busiest part of the city, especially at this time of day. Further along the path, towards the main square, was where the richer people lived. They did business occasionally for the queen. Nobody knew what business exactly, and nobody really cared.

"Why do you talk to that creepy old lady?" Jasira scoffed. "She's crazy, everybody knows that."

"Well, not everybody knows that she's nicer than she seems."

"She's too crazy for me to talk to her."

"She knew my mother. Maybe, there's a part of Prianaj that reminds me of her," Vaika said.

"Oh, I didn't know."

They crossed the famous railway underpass to the infirmary for the Procedure. Three girls were waiting outside, presumably for the same reason as Jasira. Two sets of royal Cerberi marched around the front of the infirmary with their batons swinging in motion to their steps. They wore all-black suits, and the only skin showing was above their necks. Even then, they had a metal mask that covered the bottom half of their faces, much like a muzzle. Each had a different letter marked on their chest, which was always a different colour. The letters were small, but Jasira still noticed them.

"Why are there guards?" Jasira quivered.

"They are probably just here to make sure things go to plan today."

"Something could go wrong?" Jasira raised her voice slightly, attracting side glances from the other girls in line.

"Shush and calm down," Vaika said, trying to reassure her as she pulled Jasira to the side between two cement columns. "They do this every year in June, it's nothing special, just protocol. Do you know the prophecy that was floating around a few years ago?"

"I remember there was a prophecy but not exactly how it went."

"Okay. Well, the short version is that in the month of March, a child will be born."

"Children are born every month," Jasira said.

"This one will be different. All children born can only be female, that's the reality of this world right now, that's what's happened ever since the Culling. But the child born in March will be different. The prophecy states that it will be a boy."

Jasira turned her head and looked at the Cerberi patrolling the perimeter.

"That must be why they are here," she whispered.

"Exactly. It's June, so everyone having the Procedure this month will be examined. Nine months from now will be March."

"Why doesn't the queen just put a hold on the Procedure until the month passes?" Jasira asked.

"The Procedure is the law, and the law can only be changed with a unanimous vote from the entire royal family," Vaika began. "I think some of the royals don't want to put a hold on repopulating the planet."

The clacking from the boots of a Cerberi appeared behind them. This one was taller than the other ones.

"Everything okay here?" Her voice was slightly muted by her mask.

"Perfect," Vaika said as Jasira froze in place.

The Cerberi motioned her arm in the direction of the entrance to the infirmary. Jasira followed Vaika inside through the large double doors. The walls had once been lined with dark brown tiles, but most of them had fallen off. Sunlight entered the room through a dirty skylight above them, and directly beneath it was a round glass table with a few broken coasters, surrounded by two cracked leather couches.

Jasira went and sat down, causing dust to erupt into the air from the couch's tired fabric. The girls who had been waiting outside entered and squeezed themselves onto the second couch, completely ignoring Jasira. The girl who sat in the middle looked far too young to have the Procedure; her nose was petit, and her legs barely touched the floor.

The receptionist cleared her throat, looked up at Vaika, then the girls, then back at her paperwork. Vaika sat beside her sister, brushing off some dust from her shoulders. Jasira looked at one of the girls, who had black hair and was giggling to the side.

"Today, on your twenty-first birthday, you will undergo in-vitro fertilisation by means of bone marrow from your left femur," the receptionist said.

No wonder people just called it 'the Procedure', Jasira thought. Nobody could ever remember this part of the process.

"We shall begin in—"

One of the other girls coughed uncontrollably. The receptionist glanced at the girl for a few seconds with bulging eyes, making sure the silence was uncomfortable before continuing.

"We shall begin in alphabetical order. First, Anjelika Karlom. Follow me."

Anjelika immediately got up and followed the receptionist through the bifold doors behind the couch they sat on.

Vaika waited for the receptionist to be out of sight before speaking.

"Do you want me to stay with you until you go in?" She asked as she rested her arm on Jasira's shoulder.

"I think I should be okay. You can go see the healer if you want." Jasira paused and looked at Vaika, trying but failing to read the expression on her face. "You came with me because you wanted to see the healer, right?"

"Yes, sorry. I was focusing on you, Sira. I can wait until after your turn. No rush."

Jasira smiled.

"Make friends, talk to them," Vaika encouraged.

"No, I'm okay," Jasira insisted as she let out a shaky breath.

Jasira was shy and didn't like making friends. Perhaps this was because she relied on Vaika all the time.

"What are your names?" Vaika asked the two girls on the other couch.

"I'm Therese."

The other girl next to Therese hesitated.

"I'm Vaika, and this is my sister, Jasira," Vaika responded. "What about you?" she asked the other girl.

"Florian," she said.

Thank goodness I'm not last, Jasira thought.

"Sisters? I thought the law was one child per person," Florian asked.

"I never knew my mother." Jasira said, somehow finding the courage to speak. "Vaika's mother took me in as her own."

"Then my mother also died," Vaika said. "And I took care of Jasira as if she were my own. It's been just me and her for fifteen years."

"Oh, sorry about that," Therese said as she wiggled in her seat trying to find a comfortable position.

They sat in silence for a while, waiting anxiously for the receptionist to call another girl.

Eventually, a bell rang from across the hall, and two healers rushed from one end of the room to attend to a patient calling for assistance. Florian jumped out of her seat to see the commotion, but the receptionist snapped her fingers from behind her, signalling for her to sit back down. Nobody noticed she had returned to her desk. The receptionist called two of the Cerberi inside to watch over the girls. They stood on either side of the couch with their backs turned to Florian.

"How long has it been?" Florian asked.

"Not sure. Maybe twenty minutes," Therese said.

"Are you in a hurry?" Jasira asked, almost aggressively.

"No." Florian laughed. "I just don't want to be here."

The Cerberi turned and shushed the girls.

Jasira rested her neck on the back of the couch and turned her head to the side. She suddenly felt an overwhelming sense of boredom. All she could do was count the few remaining tiles on the walls. Her eyes followed the brown squares up and down the walls to the vase of the dahlias by the entrance.

"Those are my favourite flowers," Vaika said.

"I remember. They were your mother's as well," Jasira replied.

The receptionist rose from her chair, her shoes clacking on the tough worn-out hardwood floor. The floor had faded so much over time that nobody remembered how it looked like when it sparkled.

"Florian Vurma," she called. "Follow me."

The Cerberi closest to her moved out of the way for Florian to pass. She climbed over Therese and stumbled a bit over her own feet.

Therese's legs trembled as she twiddled with her thumbs.

"Nervous too?" Jasira asked.

Therese just nodded and looked the other way.

"I don't want to do this either," Jasira continued, flicking her hair behind her shoulders.

"Stop talking, you will get us in trouble," Therese said, looking at the Cerberi.

"Sorry, I'm just nervous. Really nervous."

"My mother wanted another child," Therese began as she crossed her legs. "But they wouldn't let her."

"This whole plan is supposed to be about repopulating the earth, not letting the human race go extinct."

Therese leaned in and covered half her mouth.

"I don't think the plan is to repopulate the earth," she whispered.

The Cerberi immediately turned around and faced Jasira. The taller one grabbed Jasira's arm and dragged her to another seat on the other side of the waiting room.

"Quiet!" the receptionist growled. Her voice reverberated through the entire building.

Jasira looked across the room, but Therese looked away from her, not wanting another incident with the Cerberi.

Vaika looked at Jasira with worried eyes and mouthed, "Sorry."

The receptionist rose from her chair and leaned against the desk. She exhaled loudly, and it seemed this was the last place she wanted to be.

"Jasira Revken," she called out. "Follow me."

Jasira's heart sank so deep it felt like it was beating in her stomach. A chill danced down her spine all the way to the very tips of her toes. Her eyes widened slightly, and her legs weakened as she tried to get up.

"You're okay," Vaika said, trying to encourage her from the other side of the room.

It happened too quickly for Jasira to process. One minute they were in the waiting room, and the next they were walking down a well-lit hallway.

"Our healer will be with you shortly," said the receptionist as she opened a steel door.

Jasira stepped through and sat on a wobbly metal chair, almost losing her balance. The room had the same brown tiles as the waiting room, although there were more of them left. It wasn't much bigger than her bedroom; with the door open, another person would barely fit inside. She tried to hold onto the sides of the chair, but her hands slipped because of her sweaty palms.

The healer, a wiry woman who looked to be in her fifties, entered the room holding a small glass and metal canisters. She placed them on the bench attached to the back wall and then closed the door loudly and, with a poker face, scrutinised Jasira's body, scrunching her nose twice as she did so.

"What do they call you?" the healer asked.

"Jasira. Jasira Revken." She forced the words out.

"Jasira. Not a common name around here."

"No. It's not."

The healer then opened the metal canisters and began to pour the liquid from one canister to the other, shaking them vigorously to create a mixture.

"What's that?" Jasira asked.

"Genetically engineered proteins that will collect your bone marrow for reproduction," the healer said as she delicately poured the orange solution into a glass cup that was no bigger than a shot glass. "It's not easy to get the mixture right. Before Diektra, other cities fought for this type of research, but the monarch secured it for us. That's why it's necessary for you to drink it today. Go on, drink," she said as she handed the glass to Jasira.

Jasira took a sip and contorted her face. Its harsh sour taste trickled down her throat; some of it stayed in her mouth and began to froth. Her stomach rumbled, but it wasn't because she was hungry. She felt gasses building up suddenly in her lower abdomen.

"That's disgusting. I can't have the rest." She gagged.

"Drink," the healer repeated. "All of it."

Jasira shut her eyes firmly, held her nose and forced the mixture down her throat. She clenched her teeth as her body tried to reject it. Her head vibrated, and she nearly threw up all over the healer.

"When do I have the Procedure?" she asked as she tried to use her saliva to get rid of the aftertaste.

"Calm down." The healer laughed. "That was it."

Jasira's vision became foggy, and her consciousness slipped away. She fought to regain her consciousness, but her attempts were in vain. Her body slumped in the chair, and she went into a seizure. Jasira slipped and fell onto the cold tiles and began to tremble all over.

She struggled desperately to breathe as icy-cold water flooded over her. Her throat burned as if she had swallowed hot coals. Her body swirled around as she cried for help, but only bubbles of air came out of her mouth. She opened her eyes, but everything was a blur. Deep, red water swallowed her whole.

Jasira shot up in bed, panting hard and looking around the room frantically to make sure she was still alive. It was the first time Jasira had actually slept in a real bed, not just pieces of plastic lining and thin cardboard. The pillow felt soft on her head, and the mattress cushioned her body effortlessly. It was unfortunate she focused solely on her nightmares instead of the comfort of the bed, but it was more than a nightmare for Jasira; a part of it felt too real.

"Sira. Breathe, breathe." Vaika held onto her sister's hand and breathed with her. "Are you all right?" she asked.

"It's the second time... I've had... this dream. It just never stops." She paused, trying to keep up with her breath. "I feel like I'm in... pain and drowning... forever."

With every breath Jasira took, she felt a sharp pain in her chest, as if someone was poking it with their finger.

"It was just a dream. You're okay," Vaika said, trying to reassure her, as she tapped on Jasira's arm.

"It's not, though. It was like I was really drowning," Jasira cried.

"It's all right. I'm here, Sira. The healer said you had a seizure. Nobody has ever had a seizure after their Procedure."

Jasira felt a sharp pain in her left leg as if knives were piercing her flesh. She held onto her thigh and squeezed tightly to suppress the throbbing pain in her bones.

"Vi, it hurts." Jasira hadn't used Vaika's nickname in almost three years. "It hurts really bad."

"It means the Procedure is working. The liquid is finding healthy marrow. You'll be fine in a few days."

"Days?"

"One second, Jasira." Vaika rushed out to the corridor and stopped a healer in her tracks.

She closed the door behind her but left it ever so slightly ajar. There was a glare on the door's glass panel, but Jasira could only make out the movements of their shadows. Vaika whispered and the healer barely spoke. All Jasira could hear were quiet murmurs of their conversation. Her headache was far too disorienting, and it concealed their sentences. Jasira's curiosity got the best of her, and she painfully bent forward to catch their words. Through the glass, Jasira saw her sister cover her mouth as tears erupted from her eyes. Jasira's eyes flickered, and instantly her sister disappeared from the hallway.

CHAPTER

Two

WITHERING DAHLIA

I T WAS TIME FOR JASIRA TO LEAVE; she had already spent three days in recovery. The walk back home was strenuous, too strenuous for Jasira's body. She limped, holding onto her sister for support. The main walkway was busier than usual, so they took the backstreet. The path was a bit longer with more twists and turns, but Vaika didn't want Jasira to aggravate her leg injury.

"The healer let you out earlier than when I had the Procedure," Vaika said.

"Well, maybe I should have complained more because it still hurts," Jasira responded. "She might have let me stay longer."

Jasira sat on a rough boulder as her leg throbbed. Vaika stretched her legs and rubbed her sore shoulder where Jasira had been leaning on her. She kicked rubble up and down the alley until Jasira was ready to continue. Jasira flicked away some moss that had landed on her forearms from the building above her. She nicknamed this part of the city 'the Wetlands' because it was probably the dirtiest compared to the rest of The Border. Whenever rain fell, it became trapped in pockets of fallen buildings and soaked into the ground. These buildings overgrew with vines that broke through the ground and moss that clung to anything it could find.

They continued on their way, carefully navigating around the obstacles of fallen timber and pieces of broken fibreboard that were scattered on the ground. Getting a splinter would be the last thing they needed right now.

Once back inside their home, Jasira rested her aching body on the single, torn armchair to the side of the main room and placed her leg on a wooden stool in front of the chair.

"Vi. You haven't been yourself. What's going on?" Jasira exhaled. "Come, sit." Jasira moved her right leg off the stool and shifted her left leg to make enough space for Vaika to sit.

The room was quiet; the only sound was the low hum of nearby machinery outside. Nobody ever touched the machines; they were far too old to operate, but their batteries still worked. The machines were shaped like giant cubes and had huge vines climbing up them, much like the majority of buildings in The Border. At least once a day, the machines automatically turned on, but nobody knew exactly what they did. Jasira assumed they were generators for the city. Electricity was rare and expensive, and only the Cerberi had control of its distribution. Jasira and Vaika were never allocated any; nobody on their side of the city was.

Vaika sat uncomfortably on the stool and kept her head down; her hair hung in front of her eyes. Jasira propped herself up on one elbow and rested her head in her palm.

"The other day... when you had the Procedure... the healer gave me some news," she began, pausing after every few words.

Jasira sat back up, paying full attention to her sister.

"What news?" she asked.

Vaika lifted her head, which seemed to have been weighed down by the thick cloud of intrusive thoughts on her mind. Her eyes were slightly red but not bloodshot, and her pupils were dilated.

"The healer said it was the reason I lost the baby five years ago."

"What is it?"

Vaika lifted her shirt slightly off her back. Her skin was stained with a large, darkened spot that seemed to spread out through her surrounding veins.

"Night patch," Jasira said. "How long have you known? It looks like it has spread quite a bit."

"Six months."

"Six months!?" Jasira repeated. "You've had this for six whole months. The other day you said it was nothing."

"Probably longer," Vaika began as she put her shirt back down. "It's genetic, my mother died from it."

"Is there anything the healer can do?"

"No," she said as she sniffled, a single tear rolling down the soft skin on her cheek.

"You were diagnosed at an early stage, right?" Jasira asked.

"The healer is unsure which stage it's at yet. But they don't think I have much time left."

"How long? Did she say?" Jasira asked.

"A month. Maybe two. The healer can't say for sure."

"Why didn't you tell me earlier? You could have spoken to me about it."

"I couldn't burden you with this."

"If anything, I would be the burden to you. And you left it for months before seeing someone about it." Jasira raised her tone slightly and began to break into tears as she spoke.

"Okay, well, now I know that was a mistake. I didn't think I needed anyone to take care of me. Our whole lives, we've been fine on our own."

"I need to rest. I can't do this right now." Jasira said, ambling to her room.

Jasira rolled over in her bed, tossing and turning, but couldn't find a comfortable position. There was too much light coming into the room through the small, frosted glass window. A small section of the window was clear, which Jasira peeked through whenever she couldn't sleep. She couldn't see much, but the artificial lights from the tower opposite her room were enough to distract her. She couldn't quite figure out why, but she felt a deep sense of connection with the city's lights. Perhaps it was the fact that the lights were being controlled by someone, or something, she didn't know. The lights turn on and off automatically, like most of the things in the city. The people weren't free. Jasira knew that. But every time she looked at the lights, it made her a feel bit more powerful. She removed a crate from her pillow stack and then lowered her head onto it. Sleep overtook her as quickly and suddenly as the Culling had once taken man.

Prianaj's roosters crowed three times to announce the beginning of a new day in July. Her roosters were ridiculously loud and could be heard in the city even though she lived closer to the mountainside. Everybody on her side of the city wanted to kill those roosters of hers to keep them from losing their minds. One month later, Jasira was still distraught from the unfortunate news of Vaika's night patch. The healers still hadn't figured out a safe way to treat it or at least to stop it from spreading. Vaika constantly felt pain in her lower torso but tried to hide it whenever Jasira was around. It wasn't just physical pain Vaika tried to hide either. She masked her emotions very well under her warm smile and doe eyes.

"Sira, I have to go see the healer again. Will you be all right here?"

"Another check-up?" Jasira asked. "This is already the third one this week."

"You know there's nothing they can do. But at least if I keep going, I can pretend everything's fine." She exhaled and shrugged her shoulders. "I should be back before noon, if not, don't wait up for me. You know how they are, there's always a delay."

Vaika left through the tunnel in the wall, disappearing into the fog that lay beyond it.

Noon came around, and just as Jasira suspected, her sister hadn't yet returned home. The sunlight directly hit the makeshift aluminium roof, heating up the entire house.

The only thing that made the heat bearable was the occasional current of cool air from the tunnel. Over the next few hours, Jasira became restless, constantly moving in the armchair and wiping the sweat off her forehead. If Vaika hadn't told her to stay home, she would have marched to the infirmary to find out why Vaika was taking so long.

Jasira heard a loud knock on the wooden door behind her. She and Vaika had long stopped using that door because of the large pile of junk blocking it. Jasira hobbled toward the door, throwing the rubbish boxes and fabrics to the side. The brass doorknob felt warm in her hands and wobbled as if it was going to fall off at any moment. A rush of air swirled the dust outside. Jasira coughed and fanned away the dust around her eyes until it cleared.

It was an old friend of Vaika's, Alixem. She was a bit older than Jasira but not old enough to carry most of the responsibility for her household. As children, they played whenever and wherever they could, but as time progressed, they drifted apart. Alixem lived closer to the centre of the city, and her mother didn't like it when she strayed from the wealthier side. Jasira didn't greet Alixem; it felt too awkward. She just stood motionless and blocked the entrance.

"It's Vaika. Something's wrong," Alixem said, not greeting Jasira either.

Jasira burst out of the building, pushing past Alixem, and then slammed the door behind her.

"What's wrong with her?" Jasira asked.

"They didn't tell me. They only wanted me to bring you to her," Alixem began. "But," she paused, "it doesn't look good."

The news ignited a burning fire inside Jasira, and she began to storm towards the infirmary. Alixem followed closely behind, her red hair swinging from side to side with each step. She wore the same boots as Jasira; everyone in The Border did. There wasn't really any other type of footwear in the city.

"Are we going to talk about how we haven't seen each other in months?" Alixem asked.

"I'd rather let bygones be bygones. I've got nothing to say. I just want to make sure Vaika is all right," Jasira responded.

They were so concerned about getting to Vaika that journey to the infirmary felt much shorter than it usually did. The front of the infirmary was still being patrolled by the Cerberi, but Jasira wasn't sure if they were the same ones that were on duty during her Procedure. However, as she entered the infirmary, she did recognise one of them, the tall one, the one who moved her away from Therese.

"I'm here to see my sister. Step-sister, adopted sister, I'm adopted, my non-biological sister," Jasira rambled on, emphasising that she and Vaika were not blood related.

"Name?" the receptionist asked without making eye contact.

"My name is Jasira Revken."

The receptionist sighed and rolled her eyes.

"Your sister's name?" she asked, this time looking directly at Jasira.

"Her name is Vaika Revken."

"Room twenty-nine, down the hall, fourth room on the right," she said.

Jasira and Alixem speed-walked to Vaika's room. Alixem quietly turned to enter the room opposite Vaika's.

"Hey," Jasira said, stopping Alixem. "Thanks." She smiled.

"Any time," Alixem replied. "I was visiting a friend here anyway. There was no other way the healer would have been able to reach you. I was just in the right place at the right time."

"Well, thank you. I'll see you around then."

"For sure."

Vaika lay almost peacefully on the bed if not for her wheezing breath and pale face. Her room had a giant wall-sized window that faced the back of another building. Her brown eyes caught the flickering of the ceiling light and Jasira's reflection.

"Vi."

Vaika just shook her head, looked out the window and held in her tears. Jasira sat towards the bottom end of the bed.

Again, she couldn't enjoy the comfort of the soft mattress. She was too focused on Vaika. Vaika never wanted her sister to see her this way. She was supposed to be the one to protect Jasira, not the other way around, and that wasn't supposed to change.

"Vi. What's up?" Jasira asked, rubbing her hand on her sister's leg.

Even though Jasira wished it wasn't the night patch, she knew it was what had put her sister in that bed.

"It's worse. The healer said I don't have much time left." Vaika responded.

"How long?"

"Not long." Vaika shook her head.

"You don't deserve this, Vi," Jasira whispered.

The healer walked in, turning over a stack of pages bound together.

"Jasira. Alixem reached you in time," she said.

"What can I do to help?" Jasira asked.

"Nothing I'm afraid. The night patch has spread far too quickly than any of us could have anticipated. There may not be much time left."

"There must be something we can do!" Jasira said forcefully.

"Sira, enough!"

"No Vi. There has to be something we can do. There has to be something the healers in this place can do to fix you."

Jasira had more to say, but she felt something blocking the words from coming out of her throat. She held her lower abdomen, applying pressure to mask the pain. It's too early for contractions. The thought ran through her mind. The healer looked at Jasira weirdly, not even attempting to assist her.

"Sira?" Vaika slightly shook her sister's shoulder. "Sira, are you okay? Can you hear me?" Vaika's voice faded with an echo.

Jasira felt dizzy; her eyes rolled back, and she blacked out on the bed next to her sister.

Jasira struggled desperately to breathe as icy-cold water flooded over her. Her throat burned as if she had swallowed hot coals. Her body swirled around as she cried for help, but only bubbles of air came out of her mouth. She opened her eyes, but everything was a blur. Deep, red water swallowed her whole. The dream felt real, again.

Six months went by and Vaika was still being treated by the healers, who were unsure how long she had left. Jasira's baby bump was weighing her down, and without her sister at home, she found it increasingly difficult to take care of herself. She didn't know how much longer she could hold on, but with passing every day, she wondered if it would be her sister's last. A possibility she didn't really want to think about.

Vaika's health deteriorated, and her skin became many shades paler than the last time Jasira saw her. She had lost the majority of her muscle mass and looked so thin and fragile that her bones poked out from beneath her skin. Her hair was always drenched in sweat, and her eyes puffed up like balloons.

"You don't look too good," Jasira said, trying to lighten up the mood.

"You've gotten big."

Jasira turned to the side and rubbed her belly.

"I brought you something." She pulled out Vaika's wooden lion from her back pocket and placed it gently on the bed as if it were the part of her she had been missing.

"Sira. I haven't seen this in so long," Vaika began.

Vaika placed the toy on the side table, propping it against the glass cup.

"Sira, if I'm not here when the baby is born—"

"Don't say things like that," Jasira interjected. "You're going to make it through." Jasira was trying to convince herself as much as her sister.

"I'm dying, Jasira. There's no way I'm getting any better." Vaika used her sister's full name whenever she spoke seriously. Jasira hated it; she knew she was either in trouble or Vaika was.

Suddenly, Jasira started feeling the pain in her lower abdomen again. She felt pain radiate through her torso and lower back. She felt increasing pressure in her pelvis, and a headache started to penetrate her skull, piercing her brain. She held the bottom of her baby bump with her right hand and rested her left hand on her back.

"Vi, what's happening?" she asked, almost yelling out in pain.

"I have no idea, I've never seen this before," Vaika responded.

"It's been happening all the time," Jasira began. "The first time six months ago." She sat on the bed, and massaged her pelvis as the pain eased.

"When you had an episode the first time, I assumed it was just normal. But the more I think about it, now that you're having frequent episodes, something is very different with you. Nobody else has these types of cramps, not even me when I was pregnant."

"Should I get the healer to have a look at me?"

"No!" Vaika screamed.

Jasira got up and shut the door, making sure no one was around.

"I couldn't stop thinking about... the prophecy." Vaika continued. "Your child is due sometime in March."

"Not this stupid prophecy again. What are you saying, Vi?" Jasira scoffed as she rolled her eyes.

"I don't know, nobody really knows. But you need to prepare for the worst."

Jasira drew her eyebrows together into a slight frown.

"The worst?" She asked as lines formed between her eyebrows and tilted her head.

"Look at what we know so far, Jasira. Your baby is due in March, and you have been having these weird cramps that no one else has had during their pregnancy. The prophecy says that the boy will be born in March. Just think about it for a second," Vaika replied.

It took a few minutes for Jasira to process the information. She looked towards the ceiling and then paced up and down the room, revisiting the facts in her mind.

"Okay then, let's assume the worst," she began. "What do I do?"

"First, you need to get out of the city, at least until you're certain the child is female. If anybody finds out or even suspects anything strange, you could be in danger. It could jeopardise the baby's life as well."

"All right. What else?"

"Prianaj lives just outside the city, somewhere in the mountains. She can help take care of the baby. You'll need to find her."

"Seriously? The crazy lady?"

"Jasira, I am serious. She's not as crazy as you think."

"Are you coming with me?" Jasira asked.

"And be deadweight? No, you'll have to go alone."

"Okay. I'll need to return here at least to see how you're doing."

"No. You can't return to the city until the baby is born. The Cerberi will need to archive the baby's birth. Once you're back in the city, they'll be looking for you. It's three months, you'll be fine."

"And what if my child is a boy?" she asked, anxiously.

"Then you won't be able to return, ever. And promise me you will protect him."

"Protect a male? He could be a threat."

"Promise me, Jasira. If you raise him well, the threat won't come from him."

"All right, I promise I'll keep my baby safe, boy or girl."

They heard echoes of doors shutting, and the footsteps of healers reverberated through the hallway.

"Go, now," Vaika insisted.

Jasira threw her arms around her sister's feeble body and felt her warm breath beside her ear.

"This isn't goodbye," she whispered.

Jasira didn't want this to be the last time she saw her sister either. Vaika had taken Jasira in as her own sister and cared for her deeply and unconditionally. Jasira could never leave her sister to die alone. She didn't want it to feel like a goodbye, but for Jasira, it did.

CHAPTER

THREE

THE CRAZY LADY'S CAVE

JASIRA HURLED EMPTY BOXES AROUND THE ROOM, searching for the absolute essentials to take with her. She shoved anything useful she could find into an old backpack Vaika had found in the ruins years ago. It was big enough to carry a short rope, a dagger and two empty canisters she could fill with water. She also packed a few tins of food they had stocked up on – mostly cans of expired tuna and beans. Jasira wasn't even sure if they were still edible, but she'd rather take the chance than eat Prianaj's food.

She continued tossing everything all over the house, stumbling across childhood memories. A small wooden doll she used to play with fit perfectly inside her palm. Its leg was broken in half, and it had a crack running through its side. She used to fight with Vaika to play with it. That was until Vaika carved her wooden lion.

Jasira found some thin blankets and spare clothes to take with her. Stacked in a small pile of books was the pocket Bible her sister had gifted her six months ago. She placed it inside her back pocket to keep forever as a memento of Vaika.

Someone pounded at the back door and an unknown voice yelled Jasira's name. Then the house shook with more pounding on the other walls. She immediately raced to throw any last-minute items into her bag and headed through the tunnel. With her baby bump, she didn't fit easily, but scraping her hands and knees on the metal surface as she wriggled through, she finally broke free into the outer city.

It was December, and the winter air was the coldest it had ever been. Tiny amounts of snow were falling in parts of the city where snow hadn't fallen for years. Jasira didn't have many layers of clothing to wear. She had the same boots on her feet that she always wore and thick socks that probably hadn't been washed for a while. Her thick jacket covered her entire torso, with almost no room for a ripped scarf around her neck.

Some Cerberi came out of nowhere, walking in pairs towards Jasira. Without hesitation, she ran as far as her legs could take her. As she ran, the terrain began to change under her feet; the pavement and cement of the city began to give way to the tall wild grass that submerged it. She wasn't used to this terrain, and neither were the Cerberi, but she knew they wouldn't be able to catch her. The city faded away into the distance as she ran further. She passed Prianaj's wooden cart, but it was empty; no fruits or vegetables filled it up like usual.

Jasira's stomach rumbled, and her tongue was parched. She tried to drink her saliva to quench her thirst, but it wasn't helping. She passed a garden that stretched quite far, most likely owned by Prianaj. The sound of running water could be heard further up the mountain. There was a flowing stream there. From what Jasira learned when she was younger, fresh water could be found in springs in the mountains. She figured it was probably where everyone else in The Border got their drinking water from. The water was clear enough for Jasira to see the pebbles at the bottom. She scooped her hands together, and they almost froze. She didn't care, though; her throat was as parched as the desert in the summer. She filled her two canisters and continued up the mountain, walking alongside the stream.

Jasira kept making her way up the mountainside, getting further from The Border. The sun was beginning to set, and she didn't have anywhere to rest. She was physically exhausted as if the life had been completely drained from her body. She couldn't bear it emotionally. Finally, she saw an opening in the mountain ahead of her.

A cave, she thought.

Jasira entered cautiously, tiptoeing on the grass and hay that covered the ground. She looked around the cave, admiring how the rocky walls glowed like soft embers in the warm light from the setting sun outside. The cave wasn't large. There was just enough space for a makeshift bed at the back, and a table to the side with a pan and a short stack of metal cups.

It's strange that all these things are here. Jasira thought.

Underneath the table sat a small basket with some loaves of white and multigrain bread. Their scent wafted into Jasira's nose, and her stomach rumbled again. She crouched down and reached for a small loaf, picking up one of the multigrain buns. The crust crackled as she tore it in half effortlessly. She held it close to her nose, smelling the sweet, soft crumb. The bread made her mouth water, and she tore off a mouthful. It was as if it simply melted in her mouth, disappearing past her teeth.

"You shouldn't eat the first thing you see. I could have poisoned it." A voice echoed on the stone walls. The deep, clear voice was familiar.

Jasira gulped, swallowing the rest of the bread in her mouth, and turned her head to the entrance. Silhouetted against the sunlight was the woman who had just spoken. Jasira squinted her eyes and tried to block the sunlight from entering her eyes. The woman's bushy hair covered most of her face, and she wore thick layers of clothing. Jasira recognised who it was.

"Prianaj!" Jasira breathed a sigh of relief. "Thank goodness."

"Jasira, isn't it?" Prianaj rolled Jasira's name off her tongue. "Why are you here?" she asked, taking off her sheepskin jacket and prepared a fire.

"I left the main city."

"Yes, I can see that." Prianaj said, pointing to Jasira's bag of supplies. "I saw your sister today."

"So you should know why I am here." Jasira assumed.

"Yes, I was expecting you."

"So why ask?"

"To see if you know what you're doing here." Prianaj emphasised.

Jasira stepped back on the dried hay that crunched under her feet.

"There's no need to be afraid, Jasira, your sister trusted me enough to take care of you and the baby, or else she wouldn't have told you to find me. How long do you plan on staying?"

"Only three months, until the baby is born. Then I'll return to the city."

"If you return to the city when the baby is born, they'll ask where you've been for the last three months. And you can't go back if the baby is, well, a boy," Prianaj said.

"My child most definitely is female. The prophecy can't be about me."

"You can't be entirely sure, Jasira. Do you know the entire prophecy?" Prianaj asked, taking a seat on the bed.

"Of course I do."

Prianaj rearranged some twigs in a stone fire pit as darkness slowly crept into the cave. She looked at Jasira and remained quiet, awaiting a response.

"The prophecy is that a boy will be born in the third month of the year during a blood moon," Jasira said confidently.

Prianaj still kept quiet, waiting for more.

"You're halfway there," Prianaj began as she struck two stones together, lighting the twigs and dry leaves. "A boy of royal blood will be born in the third month of the year. Immediately after his birth, a full moon as red as blood will signal his arrival." She wafted air into the fire to fuel it. "He shall take his place as leader, the first boy since the Culling."

Jasira sighed in relief and rested her legs on a pile of hay.

"I'm not of royal blood," she said.

"Your sister told me you were adopted, yes?"

"That's true." She paused, rubbing her lower back. "But Vaika would have told me who my mother was. And what if you got it wrong, the prophecy, perhaps the month is off."

"The prophecy cannot be wrong." Prianaj stressed her words.

"How can you be sure?" Jasira asked.

"It cannot be wrong because I am the oracle who prophesied it."

With those words, the sun sank beneath the horizon, and darkness completely filled the cave.

Jasira awoke to an empty cave, the spring breeze blowing inside. Prianaj had gone to the city to sell her supplies. The cave was always darkest in the mornings, as the sunlight shone on the other side of the mountain first. Jasira waited in bed, not wanting to move; her baby bump was too heavy.

Prianaj came back early with a torch in hand; its flames licked at dust particles in the air and spat them back out in tiny amber sparks.

"You're back early. Any luck selling today?" Jasira asked.

"No," she said. "I rushed back as fast as I could."

Jasira sat up, all her attention on Prianaj's words.

"What is it?"

Prianaj gently rested the torch in the holder attached to the wall. Its light, combined with the sun's, was enough to illuminate the dried tears on Prianaj's coffee-coloured skin. Her wrinkles weren't as prominent as they usually were, but her eyes were dark around the edges.

"Jasira," she began, "Vaika was the kindest soul I ever met."

Was? Jasira thought.

"She lived a lot longer than the healers predicted. Even longer than I predicted," Prianaj continued. "This morning, I went to see her. She took her very last breath in my arms. I'm so sorry, Jasira. She's gone."

Jasira felt a wave of sadness wash over her. Even though there was sufficient lighting in the cave, it felt like the light had died out. Her eyes blurred and her head became heavy. A sob rose in her throat, and a sting pierced her chest as her heart twisted. Her eyes prickled with tears, but they didn't seem to fall. She rounded her arms around Prianaj, who comforted her. It was the first time in three months Jasira had shown an ounce of emotion to Prianaj.

"I know," Prianaj said, gently patting Jasira's head.

"I didn't get to say goodbye." She choked between sniffles.

"I know what it's like. My baby was taken from me when she was young."

"How did you deal with your baby's death?" Jasira asked, using her shirt to wipe the tears from her face.

"My baby was stolen. I searched for years, but nothing. I still have no closure," Prianaj said.

Later in the day, Jasira helped pick Vaika's favourite flower, dahlia. She only picked the bright orange ones, her sister's favourite colour. She gathered some smaller flowers with large leaves as well and bound them into a wreath. Prianaj threw a black sheepskin coat over her own shoulders. She tied her rusty greying brunette hair in a high bun.

"I feel like I knew she didn't have much time left, but at the same time, I can't believe she passed so quickly," Jasira said, handing the wreath to Prianaj.

"The funeral will be rushed, but if we don't lay her to rest today, the Cerberi will dispose of her," Prianaj said as she tied the front of her coat together. "You know you won't be able to enter the city until—"

"I know, but I can be there in spirit," Jasira interjected.

Prianaj nodded and headed for the city with the bottom of her coat dragging on the floor. Jasira felt a contraction but was so focused on the death of her sister that she didn't want to say anything. She waited for Prianaj to disappear around the side of the mountain and then listlessly climbed back into the cave, using one hand to hold the bottom of her baby bump while placing the other on the ground to keep her from losing her balance. The afternoon light hit the back of Jasira's neck, almost tickling her skin. She scratched it to relieve the uncomfortable feeling, taking her arm off the ground and wobbling a bit. As she clambered up the mountain, she occasionally slipped, catching leaves and twigs in her shirt, but the terrain wasn't too steep and she didn't fall very far.

Jasira reached the cave and lay flat on her back on the hay floor. The contractions were stronger this time, and she didn't know what to do. She called out for Prianaj, but it was no use. There was no one around. Her eyelids were heavy, but she couldn't sleep now, especially with the pain she was feeling. Through her blurred vision, she saw the hazy silhouette of the mountains on the other side of the valley. She was so utterly exhausted, with every fibre and sinew in her body compelling her to sleep, that somehow the pain disappeared. But just as Jasira was dozing off, another painful contraction forced her wide awake. She cried in pain, clenching piles of hay in her fists.

Hours passed and the contractions became much stronger and closer together. Jasira felt her lower abdomen expand with pressure and was sure she was about to explode. Just then, Prianaj entered with her torch in hand and sprinted to light the cave. She pushed some trivial items out of the way and dragged Jasira by the arms across the cave onto the bed, which was much softer than the hard ground.

"Jasira, what is it?" Prianaj asked aloud.

"It hurts." Jasira's eyes rolled around, and her head swung from side to side.

Prianaj ripped open the bag of supplies she brought last month and emptied it beside her. It was filled with bandages, solutions and ointments. Most of it she made herself and traded them in the city, but she preferred the organic versions that she said were much healthier. Jasira squeezed her belly tighter and kept her legs crossed.

"Jasira. The baby is coming," Prianaj said as she placed a thin blanket over Jasira's lower half.

"No, no it can't be yet."

"Breath, in and out. Quick breaths." Prianaj emptied clean water into a bucket while performing breathing exercises with Jasira.

Jasira's screams pierced the walls of the cave, so Prianaj scrunched some fabric into a ball and told Jasira to bite down on it. The cloth was supposed to mute her screams, but somehow, they got even louder. Her cries were so loud the two of them couldn't even hear the howling winds outside.

"Jasira dear, you're going to have to push."

Prianaj waited with her hands ready to deliver the baby. The sheets were getting soaked with blood. Prianaj glanced below, her eyes tight and worried.

"This isn't normal," Prianaj muttered.

"What?"

"Nothing, you're doing great, Jasira."

Jasira lifted her head to take a look, but Prianaj signalled with her hand to lay back down.

"I can see the baby."

She splayed out her frail fingers to cradle the baby's head.

"Almost there, Jasira, just keep pushing."

The baby's head was completely out; its soft cries were music to Jasira's ears.

"My child," Jasira said, elated.

The rest of the baby came out easily and landed in Prianaj's palms. She swaddled it in a clean blanket and rocked gently. She then used the other towels to wipe away the blood and other fluids all over the baby, and then cut away the placenta with a knife before tying it into a knot.

Jasira was too far inside the cave to check if the moon outside was blood red. Her vision was blurry from the exertion anyway, so even if she had reached the entrance, she wouldn't have been able to see the moon.

"Prianaj," Jasira began as she swallowed hard, "tell me the baby is a girl."

Prianaj walked to the entrance with the baby in her arms. "Blood moon," she mumbled, looking into the sky.

"What?"

Prianaj walked back to Jasira, not knowing whether to smile or cry. She reached for Jasira's hand and squeezed it.

"Prianaj," Jasira repeated, this time swallowing harder. "Tell me it's a girl."

Her mind was reeling with the different possibilities that might be ahead of her. It felt as if she was trying to force pieces to fit together, but she wasn't completing the right puzzle. Prianaj shook her head slowly and handed Jasira her child.

"Boy," she whispered.

CHAPTER

FOUR

LETTER TO A STRANGER

J ASIRA ROCKED HER BABY BOY TO SLEEP in her arms and
then placed him gently in the circular cradle of hay beside
the bed. She had never seen a boy before; nobody had in two
hundred years, and even though to her all babies looked the
same, this baby was hers. She sat by the cradle, staring at his tiny
nose and angelic face. She stroked her finger across his smooth,
soft skin and then let the boy hold onto it as he slept. He let out
tiny noises as he yawned and rested with a smile.

How can man be so cruel? Jasira thought.

For her entire life, the people of Diektra were taught that
humanity's destruction was the result of man's greed. But she

stopped believing that after the birth of her baby boy. She couldn't think negatively about him.

Jasira gazed at her baby adoringly while singing a song that Vaika used to sing to her at bedtime when they were young. She held in her tears as she sang, remembering how her sister's voice used to soothe her.

'Close your eyes and lay down,
Dream a very sweet dream.
I'll be right here to stay,
You'll be safe when you're with me.

Close your eyes my child,
In you, I'll find my strength.
I love your beautiful smile,
I'll be yours until the end.'

Jasira's voice wasn't what people call 'show stopping', but it was good enough to put the baby to sleep. Prianaj came back from the city with a basket of spices and trinkets.

"Good morning, princess. Nice voice," she said.

"Please don't call me that. I'm not royalty," Jasira replied.

"Your baby fulfils the prophecy."

"So?" Jasira kept silent, not wanting to acknowledge the significance of what Prianaj had said.

"Of royal blood," Prianaj whispered.

"That doesn't prove anything," Jasira began. "And just because I was adopted and didn't know my mother doesn't mean I am of royal blood."

Jasira covered her baby in her own blanket, shielding him from the cool draught that entered the cave. The baby awoke and began crying. Prianaj unpacked her things and handed Jasira a small glass bottle of oil.

"Rub this on his feet every night, he'll sleep better."

"Thanks."

"What's his name?" Prianaj asked as she unpacked the rest of the basket.

"I don't know many good names, especially male ones. I'd planned to name the baby Vaika if it was a girl, but that name is no longer an option."

"Well, you better come up with something fast, you don't want to be calling him 'baby' for the rest of your life," Prianaj began as she ruffled through the basket. "I have to run back to the city. I forgot to grab some more towels. You should be fine with 'baby'."

"As long as you don't take too long," Jasira said.

"That wasn't a question."

Jasira mashed pumpkins and sliced tiny pieces of carrot in a metal bowl for when her baby awoke. She poured the baby food into some metal canisters with rubber nozzles that Prianaj must have kept from when she had her own child. Her baby boy disliked being breastfed, so liquefying produce was the only option. Jasira's heart melted, and a warmth spread through her body. Without Vaika's warning, she would have fallen into grave danger long ago.

"Thanks Vi," she whispered under her breath.

Prianaj had a small pile of ancient books, which Jasira flicked through to find potential baby names. Names such as Oliver, William and Henry popped up in multiple books, but she was looking for a name that felt more meaningful to her. In the last book Jasira picked up, there was a short handwritten letter on the first page. Jasira hadn't learnt to read very well, but she did her best to read aloud the legible words. She thought it was bizarre that the recipient's name was smudged out completely.

Dearest ,

My child, although I may never meet you,
know that I love you. Forever in this world
and forever in the next.

Your mother,
Prianaj

Jasira couldn't understand why the name was smudged out. If Prianaj was never going to meet her child, why would she write a letter? She tore out the page, closed the book slowly and then hid it underneath the rest of the pile. She folded the letter and slipped it underneath the baby's cradle but accidentally lifted the cradle too high, waking the boy from his nap.

"No, please stop crying," she said to him as if he could understand her. "Shush, child."

Jasira gave him a bottle of vegetable mash, and he immediately stopped crying.

"Look at me go, this doesn't seem too hard."

She spoke too soon.

A foul odour wafted into Jasira's nose, and she wished she didn't have to deal with it.

"Eugh!" she whined, closing her nostrils with her fingers.

"Oh lord, the smell!" A voice screamed.

Jasira smiled from ear to ear, and relief flooded her body.

"Prianaj. Thank God you are here," she began. "He shat himself, I don't know what to do."

"You're lucky I've got more towels. Fill the bucket with water, he needs a bath." Prianaj said, closing her nostrils as she reached for the child.

As Prianaj lifted the baby, the cradle shifted, revealing the letter underneath, but she was too preoccupied with the foul smell to see it. Jasira heaved the bucket into the cave and left it at the entrance in a hurry to learn how to change her child. The corner of her eye caught the exposed beige paper by the cradle. She tried to manoeuvre her body to the other side of the bed while still looking at the baby so Prianaj wouldn't notice the letter.

"Did you learn to do this with your child?" she asked, trying to find out more about the letter.

"My baby was stolen, remember?" Prianaj said bluntly.

Jasira quickly picked up the letter from the floor and shoved it into her pocket. Prianaj turned and looked at her weirdly.

"No, but before your baby was stolen, didn't you get any time together?" Jasira asked.

"Why are you all of a sudden concerned with my past. We have bigger problems to worry about."

"What problems?"

"The women in the city saw the blood moon the other night," Prianaj began as she lowered her voice. "They know a boy has been born."

Prianaj threw a clean towel over her shoulder, took the baby and bathed him in the bucket.

"The blood moon could be anything. They don't know a boy has been born."

"But word has spread to most parts of The Border. When the queen finds out, she'll send Cerberi to both cities. It won't be long before they start looking here."

The other city, Monday, was on the other side of the mountain range, closer to the Pacific Ocean. It was a three-day walk from The Border. The queen, Samaari Weiluk, ruled both cities from her citadel, whose location was secret.

Jasira's thoughts made her heart beat faster. She couldn't process what Prianaj was telling her.

"I have to protect him. If that means we must flee from here, then we run."

"Where to?" Prianaj asked, wrapping the baby in the clean towel.

"Monday. If the Cerberi check there first and don't find anything, then they'll come here."

"That's at least a three-day walk, maybe four with a baby. We stay here, for now."

The wind blew across a stone bridge that connected the citadel to the mountain. The roaring waterfall, opposite the bridge, sprayed water all over the Cerberi running across it. Samaari sat on her metal throne; her sister, Elaine, stood beside her.

"Idira," Samaari said, acknowledging her messenger.

"Your Majesty. Forgive me for interrupting you like this, but I have some news from the cities," Idira said.

"What news do you bring?" Samaari asked, signalling her to continue.

"The people believe... they believe the boy has been born."

"Our astrologers haven't observed a blood moon," Elaine said.

Samaari glanced at Elaine; her piercing silver eyes stared deep into her soul. Even in the dimly lit room, Idira could see the fear in Elaine's face. Elaine stepped to the side and kept her head down.

"If I didn't promise Mother I'd take care of you when she broke the law to have you, I would have cut your tongue out years ago," Samaari warned. "But you are right. Our astrologers haven't observed a blood moon."

"Forgive me then," Idira began. "It must just be rumours floating around."

"Do you know what happens when people start talking? When rumours surface?" Samaari asked as she slowly rose from her throne.

"No." Idira gulped.

Samaari stepped down to the lower platform, dragging her bare feet on the stone-cold floor.

"All of a sudden, they believe they have power. But they are the inferior ones!" she growled.

Idira nervously exhaled a small misty cloud of breath into the cold air of the citadel.

"Apologies, your majesty. The people must be mistaken."

"Mistaken they are. Fix it. And the rumours, squash them," Samaari said slowly.

"I will send Cerberi to both cities."

Idira hurried out of the throne room, slamming the door on the way out.

"People are delusional," Samaari scoffed as she regained her seat. "They actually believe a boy was born."

"And if the boy was born?" Elaine asked, hiding the fear in her voice.

"No need to worry sister, he wasn't," Samaari said reassuringly, lifting her legs over the armrests. "But if the boy was born, I'd make sure he doesn't take another breath."

"You would take the life of a baby?"

"A baby who would grow up to believe he's better than all of us. And you know what he'd do?"

Elaine shook her head, widening her eyes.

"Take over and infect this world like a virus," Samaari whispered. "I'd be doing the world a favour. If the Culling didn't wipe out the men, they would have murdered each other. Oh right, they did."

"Maybe a boy would be born for a reason."

"Are you defending man? You know what men are like," Samaari said, shivering in disgust.

"Forgive me, sister, but no, I don't. Nobody has for two hundred years. We only know what we learnt about them. You wouldn't know what men are like either."

"You're going crazy, Elaine. Take a break, go for a walk, take a bath, rid your mind of those toxic thoughts."

"Don't condescend to me."

"I'm the older one. I'm the one who had to silently watch as our grandmother was murdered for breaking the one law our founders set. You're lucky you didn't have to bear it. Not to mention our own mother for breaking that same law."

"Are you not the one who banished mother's sister? If only I could have changed that law, but we are missing the third piece!" Elaine yelled.

"What do you think gives you power? Is it your crown? Is it the fact that you are queen?"

"My blood."

"The same royal blood that flows through my veins right?" Elaine said.

"I'm not doing this with you," Samaari said as she stormed out of the room.

Elaine hurried to exit the citadel, leaving behind everything she held dear: her materialistic belongings, her sister and, more importantly, her status. She slipped past the Cerberi and took one of the regal horses to ride to the closest city, Monday. She laughed as joyful tears came to her eyes. For so long, Elaine had felt trapped under her sister's command, unable to break free, and she didn't expect it to be this easy to leave. It took a long while for anybody to notice she was missing but Elaine wondered if they would even care.

Prianaj cleaned up the mess she made while cleaning the baby. Jasira sat at the edge of a cliff, dangling her feet as she read Prianaj's letter repeatedly. The landscape opposite the cave gave her a sense of freedom. Forests bloomed and overgrew the ruins of pre-war cities. Flocks of birds flew by, and their chirps echoed through the quiet woods.

"My bones are old," Prianaj grunted as she tried to sit.

"You gave me a fright. How's he doing?" Jasira gasped, hiding the letter under her leg.

"He's asleep. He's a beautiful child, Jasira."

"I don't deserve him."

"Don't say that. I've only known you for about three months now, but I can certainly say that he deserves you."

"Not if it means he could be killed because of me. I don't want that. He doesn't deserve that."

Prianaj smiled and tightened her jacket around her shoulders. Jasira thought for a second about how to question

Prianaj concerning the letter without seeming too suspicious. Her mind raced with questions to ask.

"If you fear for his life, then he does deserve you," Prianaj repeated.

"Forever in this world and forever in the next," Jasira said too quickly.

"What did you say?"

Jasira didn't think it through properly.

"It's just something my sister used to say. Apparently, her mother used to say it to her."

"No, she didn't," blurted Prianaj.

"How well did you know Vaika's mother?" Jasira asked.

Prianaj locked eyes with Jasira and pursed her lips. The seconds that passed felt like hours.

What was I thinking? I can't get out of this now. Hopefully, that was convincing enough, Jasira thought.

"I wanted to say that to my baby," Prianaj began as she breathed in the fresh mountain air. "But I didn't even get the chance to name my child."

That's probably why the name was smudged out. She thought.

"Did you have anything in mind?" Jasira asked, swinging her legs in small circles.

"It wasn't so much the name, more the meaning. I wanted a name that represented something strong and fiery," she said, drying a single tear with her jacket. "Have you come up with anything for the boy?"

"A few names popped into my head when you were out."

"And?"

"Nothing yet that I like. Boy names are hard. All we've known are girl names."

Prianaj and Jasira spoke for a while before they realised it was getting dark. As they headed inside the cave, Jasira tripped and almost fell but Prianaj caught her just in time. Suddenly, the same nightmare Jasira had had about drowning struck her like a lightning bolt from a cloudless sky. But this time she was awake. Prianaj carried her to the bed, as Jasira couldn't move her limbs until the nightmare ended.

"Calm down, Jasira. It'll pass," Prianaj said as Jasira slipped in and out of consciousness.

Jasira slept on the side of the bed closest to her baby and furthest from the entrance of the cave. She awoke in the middle of the night to the boy crying. The glowing embers were just enough for her eyes to catch sight of the Bible Vaika had gifted her. For some reason, something clicked in her mind, as though a light bulb had gone off in her head.

"Adam," she whispered. "That's what I'll call you. Adam."

CHAPTER

FIVE

ESCAPE TO MONDAY

THE VEGETATION IN PRIANAJ'S GARDEN grew much slower than in previous months. Jasira fertilised them with a concoction of dead plants and some water Prianaj had previously mixed together. She didn't really know if it helped or if that's what stunted their growth.

Jasira enjoyed helping out in the garden. It took her mind off the baby every once in a while, and she couldn't let Prianaj do it all by herself. She also helped to feed the animals. Prianaj only had roosters and pigs. Jasira was eager to learn since there was no garden or animals back home in The Border. Vaika always brought home freeze-dried and non-perishable food.

Some nights, they went without any food. Vaika didn't have a very good job and earned very little. The only thing that got them through the seasons was the food assistance every mother received after the birth of their child. Although Vaika was never granted it for her own child, she was given assistance for Jasira because she was her primary carer. This made it easier to get food, especially when nobody else could care for them both.

From a close distance, Jasira could see dust swirling up from the ground towards the sky. She initially assumed it was a sandstorm but soon realised it was far worse. A string of horses ridden by Cerberi, no less than fifty of them, making its way towards The Border.

Jasira ducked and hid behind an overgrown bush. She watched them ride past her, kicking dirt into the air, and stopping just in the front of the city.

"Oh no! Adam," she said to herself as she clenched her teeth.

Once the Cerberi were completely out of sight, Jasira sprinted to the cave, intermittently looking behind to check if they were still close. She saw Prianaj up ahead, who didn't seem to have noticed the string of horses.

"Prianaj," she whispered aloud. "Get down."

It took Prianaj a second to notice the dust storm created by the horses before she reacted.

"They're here for me," Jasira said, breathless and tired.

"Not you," Prianaj responded. "Him." She pointed towards Adam.

"We have to go."

"They'll look in the city, they won't find you there and then they'll leave."

"They'll come here. I'm sure of it. They'll probably ask everyone who lives in The Border, and then they'll come for you."

"Nobody's seen you in months. There's no record of Adam's birth. They'll assume you've died or run away."

"I'd rather run away than be dead," Jasira scoffed.

She began packing her essentials into her backpack. For Jasira, repacking for another journey felt familiar, a sort of déjà vu she could never escape.

"Look, Prianaj. I'm really grateful to you for willingly taking me in over the past few months and helping me take care of Adam, but I have to leave. You can stay here if you like, or you can come with us."

"And where will you go? There is nowhere to go," Prianaj insisted.

"I will find a place. I'll go to Monday. The guards would have searched there already."

"Or they might be going there after they don't find you here. They'll know you ran away."

"Let them try to find me, they'll be looking for a while."

"Go on, then. See how far you can make it alone, it's no journey for a child, and I'm not talking about the baby."

Prianaj left the cave, taking her wooden walking stick with her. From the way she walked out, Jasira knew she wouldn't see her again for a long time, maybe never. Nonetheless, Jasira continued to pack her things, more so for Adam than for herself. The more Jasira packed, the more she realised how much she actually needed Prianaj for the journey. She would never be able to take care of Adam on her own, changing him, bathing him, even putting him to sleep or calming him down.

She finished packing, but leaving this way didn't feel right to her. She couldn't bear leaving without seeing Prianaj again. After all, even though Jasira didn't initially like her, Prianaj has allowed her to stay in the cave for several months. Not many people in The Border would have done that.

Jasira waited in complete silence for Prianaj to return, not even Adam made a noise. She changed him into some clothes that Prianaj had sewn that morning so that she'd have something to remember her by. Eventually, she decided it was time and picked up her things to leave. Right at that very moment, Prianaj returned from her walk.

"You aren't thinking of leaving in complete daylight, are you? They'll catch you so fast you won't have time to blink," she said, laughing.

"Oh, Prianaj. I'm glad you're back. I didn't want to leave without saying goodbye," Jasira said thankfully.

"What makes you think you're leaving without me? I've seen you try to take care of that baby. You'll need me," she said, knowingly.

"But I thought you were staying here."

"I just needed to think it through in fresh air," Prianaj said.

"So what's your plan?" Jasira asked.

"Well, your plan wasn't too bad," she began. "We can go to Monday, but the journey will be difficult."

"How difficult?"

"Very," Prianaj emphasised. "It's like a desert out there. You'd dehydrate if you weren't prepared."

From a tiny trapdoor underneath the hay, she pulled out a hand-drawn map of the two cities and spread it open on the bed. She pointed to a mountain range just outside The Border.

"We are around... here," she said. "We need to get here." She pointed at the city labelled 'Monday'.

It was not too far from The Border, perhaps a two-day walk instead of four.

"We have to go through The Border. It'll save us time."

"Can't we just go around? I don't want to risk being caught by the Cerberi. There are going to be more of them now," Jasira recommended.

"The mountains are far too dangerous to carry a baby across. And it'll take longer, so we must go through the city. We have to leave at night, the Cerberi won't see us."

"All right then, we leave at dusk. I'm doing this to protect Adam."

"If the Cerberi find the boy, it'll be the end of all of us." She paused and looked at Jasira. "For the record, I like the name you chose for him."

The night couldn't have come fast enough. Jasira was more than ready to start the journey. They proceeded down the mountain with their path lit by the fiery torch Prianaj held in her hand. Jasira was carrying Adam in a basket covered with a blanket. Thankfully, he was asleep. Any noises would attract the attention from the Cerberi, especially noises from this baby.

"I can't believe you're making me leave behind my life's work. It took me forever to grow my garden. And those animals were not cheap," Prianaj whispered.

Jasira pretended to ignore what Prianaj said as they passed by the garden.

"You'll need to put that torch out in a minute. We're almost there."

"Actually, I was planning to leave it on so the Cerberi would see us," Prianaj joked.

Jasira rolled her eyes.

The front gates of The Border were left open as usual; no guards ever watched them. Their society had been built on a foundation of trust, but everybody who lived there knew it would never last, especially since it was already crumbling. The main walkway was completely empty. The Cerberi weren't patrolling the square as usual, and the citizens always cowered in their homes. The city lights were on, which illuminated their path, but Jasira and Prianaj stayed in the shadows for the majority of the walk. They took a sharp left turn just before the underpass and then rounded the corner on the right.

Behind a complex of restricted buildings was the one building nobody ever entered. People joked that it was haunted. Some even claimed to hear the screams of ghosts coming from inside it. Jasira didn't believe in ghosts. It was a derelict structure made of rotting wood and a stone encasing. They hadn't seen the building this close before but were hurrying past it when a young Cerberi appeared from behind. Prianaj pulled Jasira out of the light and held her against a shadowed wall. She gestured that they should move along the path towards the back exit.

"Slowly," she whispered.

Jasira must have accidentally stepped into the illuminated street because the Cerberi's footsteps grew increasingly close.

"Ma'am!" the Cerberi shouted.

"Run," Prianaj said.

"Excuse me, ma'am!" the Cerberi repeated loudly.

They zigzagged through narrow crevices between buildings, almost smashing into walls at every turn. They were only a few meters away from the back exit of the city when the Cerberi blocked their path. Jasira's eyes widened, and her mouth dropped halfway. Her heart pounded in her chest, and her muscles tightened.

Shit, we're dead now, she thought.

"You dropped this, ma'am," the Cerberi said, handing Jasira a small satchel that should have been attached to the backpack.

"Oh, thank you," Jasira exhaled, taking it back. "I didn't realise I dropped it."

"What are you two doing out this late anyway?"

"We were just taking—"

"This basket back to a friend," Prianaj interjected.

"I'll walk you there," the Cerberi began. "What did you say your name was again?"

"I didn't," Jasira said. But she realised if they were looking for her, she shouldn't reveal her real name, so she lied. "My name is Vaika."

The rough movements Jasira made with the basket was enough to wake Adam from his slumber.

His cry was slow at first and gradually grew louder. The Cerberi looked directly at the basket and then back at Jasira who tried to rock the basket to put him back to sleep.

"Ma'am, is there... a child in there?"

Prianaj tugged at Jasira to leave; the exit was within arm's reach. The Cerberi reached for her baton on the side of her belt.

"Ma'am. I'm going to need you to lift that blanket," the Cerberi said, pushing a button on the wall next to the exit.

The gate rattled as it began to close. Jasira turned and ran through, waiting for Prianaj on the other side. The Cerberi reached out for Prianaj, taking a swing at her with her baton. Prianaj dodged it and threw down a black rock that exploded into a towering cloud of black dust. It lingered there for a while, but once the smoke started to clear, the Cerberi waved away the remaining smoke, coughing as the smoke particles irritated her throat. Prianaj and Jasira disappeared into the dark, and the gate closed, locking them out.

It was too dark for them to see where they were. Nobody lived outside the city so the superiors hadn't installed any lighting there. Jasira assumed they were walking on grasslands, but with little illumination, she was unsure. She outstretched her free arm to check if there were any obstacles in her way. Prianaj occasionally got in the way, and Jasira accidentally hit her in the face at one point. Jasira felt a cold concrete wall in front of her and moved her hand across the surface. She couldn't figure out how far or high it stretched, and so she and Prianaj decided to make camp nearby and continue in the morning light. Jasira stayed awake and kept watch of the basket, worried that the Cerberi had followed them.

"Don't stress, Jasira. That gate won't open until dawn. It locks overnight," Prianaj said reassuringly.

"I've never seen them lock the gates before."

"They only do it if there's a threat. They won't want anybody leaving if they're looking for you. They would have locked the front one as well. They did that a few years ago."

"A threat?"

"You and me both."

"What was that you did back there?" Jasira asked.

"It's a little science I learned over the years," Prianaj answered as she dozed off into sleep.

CHAPTER

SIX

THE ELECTRIC FENCE

J ASIRA'S SKIN TINGLED AS SHE FELT the sun's warmth on her skin, waking her from an uncomfortable sleep. Her neck was sore, but after a few stretches, she felt refreshed and good to go. However, she couldn't find Prianaj. It was like she had disappeared into thin air.

"Prianaj!" Jasira called out.

She checked to see if Adam was in the basket; thankfully, he was.

"Prianaj!" She repeated.

Jasira got up from the ground and turned her back to the wall she had been leaning on all night. She saw what was in front of her more clearly in the sun and gasped, widening her eyes and stepping back. Tiny concrete houses were unevenly crammed in a large field. They looked old and abandoned, and vines were growing all over them. Some even had entire trees growing through their centres. Jasira quickly grabbed the basket and ran through the gaps between the houses.

"Hello. Is anyone there?" she shouted.

Panic struck Jasira as she realised the paths were all dead ends. She was stuck in a giant maze of narrow alleyways, some of which she could barely fit through. Her breath became shallow, her eyes blurred and the world turned upside down.

"Oh God, where am I?" She quivered.

She moved desperately through the maze, trying to figure a way out, but no matter which way she turned, she always ended up at a dead end.. Adam cried in his basket, and the walls seemed to close in on her. She had never been claustrophobic – she had lived in a small home for twenty-one years – but this was too far out of her comfort zone.

"Prianaj!" she cried once more.

"Jasira!" Prianaj's voice was close.

Hearing Prianaj was enough to calm Jasira down, and she followed the direction of the voice. She could also hear the rattling sound of the Cerberi opening the city gates; it sounded like cracks of thunder above their heads.

"Jasira!" Prianaj called again, much closer this time.

Jasira knew she was heading in the right direction, away from the gate and closer to Prianaj's voice. She turned corner after corner, jumping over overgrown foliage. Prianaj bumped into her from the side, and they both fell to the ground. Jasira spun her body to break the fall of the basket, painfully bruising her back. She ignored the pain, grabbed a bunch of dry leaves beside her and threw them at Prianaj.

"You left without me!" she yelled.

"No, no. I didn't." Prianaj defended. "Nature was calling, I needed to relieve myself. I saw the old buildings, got distracted and explored a bit too much."

"You scared me to death." She gently pushed Prianaj's shoulder.

"Sorry. I thought it would be a thirty-second expedition, but it looks like both of us got lost in the maze."

"What is this place anyway?"

"I think it's a part of the city people used to live in. Before the Culling, of course," Prianaj said as her eyes scanned around the buildings.

"Spooky."

They both laughed it off and proceeded through the labyrinth of buildings. Jasira finally felt the pain as she struggled to get up. She took a second to readjust before continuing to walk.

"This looks like the way through. That mountain range behind us is where the cave was," Prianaj said as she pointed at the peaks behind them and threw a bag over her shoulders.

The rest of the terrain was covered entirely by grassland and bushes. There were occasionally random patches of dirt and dead grass, and it looked like the absence of humans had allowed the earth to recover.

Back in The Border, it was mandatory for everyone to attend a history class. It wasn't like the typical history class where students are taught about ancient civilisations. These revised classes taught the people that human interaction with nature almost killed the earth. They were made to believe that men were to blame for the destruction of the world and its resources. Everyone believed it because the earth had healed in the absence of men. However, some women refused to accept that belief. They knew that the earth had recovered in part because there were fewer humans inhabiting it.

"I think that's Monday up ahead." Jasira said in delight, looking at a wide structure in the distance.

"Impossible. We've only been walking for half a day."

"Maybe the map was wrong."

"The map is definitely not to scale, but that is not Monday," Prianaj said confidently.

They stopped to rest in the shade of a pin oak tree. Prianaj took two loaves of bread out from her backpack for them to share. Jasira took out a bottle of mashed vegetables for Adam and a second flask with water. The blaring sun wasn't making their journey any easier, but Jasira made sure Adam was shaded from the sun the entire time because his skin was the most sensitive.

"What are we going to do when we get to Monday?" Jasira asked.

"I guess I'll start another garden and sell my produce in a marketplace, and we'll have to find somewhere to stay."

"Another cave?"

"Perhaps. You know Adam won't be able to go out often when he's older. It will be far too dangerous."

"We can cross that bridge when we get there," Jasira responded.

The closer they got to the structure, the more Jasira realised it wasn't in fact Monday. It was an enormous iron fence that stretched as far to the east as it did to the west. It stood over ten meters high and probably ran the same length deep. The top was covered with barbed wire, and it was labelled with several signs that read:

'WARNING ELECTRIC FENCE'

The fence was bordered by a thick layer of bare sandy soil.

"Shit. What a setback!" Jasira cussed.

"Who's that?" Prianaj asked, pointing at someone walking towards them.

"They're not wearing Cerberi uniform."

"Who else would be way out here?"

However, the woman was not alone. She was with another younger woman. Both of them had tied their blonde hair in a low ponytail and were wearing button-up shirts with blue jeans. Their leather boots were scuffed, and their skin was fairly tanned.

"You two are a long way from home," the older woman said. "What are you doing out here?"

"We could ask you two the same thing," Prianaj said.

"We noticed Cerberi coming through this fence last night," the younger woman said.

"Now what did I say about keeping your mouth shut, Braelyn." The older woman raised her voice and gave the younger woman a look of concern.

She was probably the younger woman's mother. They had the same facial structure and almost identical eyes.

"Don't mind her," the older woman said. "My name's Talani. Are you two travelling? I don't see too many people carrying baskets or backpacks around here." Talani spoke with her hands on her hips.

"Passing through," Prianaj said. "Your accent."

"Pardon me?" Talani asked.

"Your accent," she repeated. "I thought Southern accents were lost in the Culling."

"We moved out here long ago. I guess we just adapted with the environment," Braelyn said. "We've been stargazing, try'na catch that blood moon the prophecy talks about."

Talani looked at Braelyn again, this time with seriousness in her eyes.

"The blood moon passed some nights ago," Prianaj responded.

The whole time they spoke, Prianaj kept eye contact with both strangers, keeping Jasira behind her.

"Well, too bad we couldn't catch a glimpse of it. If the boy was born, he'd be worth a fortune. The Queen would pay a hefty price for him, don't ya reckon?"

Prianaj completely avoided Talani's question, fearing for the boy's life.

"You said you moved out here, I don't see any buildings," she said.

"What are you two looking for anyway?" Talani asked.

"Like I said, we are just passing through."

"What about you?" Talani asked, looking at Jasira. "Cat got your tongue?" She laughed hysterically.

Paralysed with fear, Jasira couldn't speak.

"She's mute," Prianaj said. "Cerberi cut her tongue out years ago."

"Shame. Couldn't imagine your pain."

"We best be going now."

"Whoa now, hold on. You two don't expect to find a way through the fence, do you? There's no way through, at least not for miles. Follow us, we can get you two through," Talani insisted.

"We can make it just fine on our own."

They walked past the two strangers.

"What's the matter, don't trust us?" Talani began. "Look, this fence isn't even electrified."

Talani touched the fence with her bare hands. Prianaj was expecting a reaction, but Talani was right; the fence was in fact not electrified.

"That was stupid."

"I can help you two get across."

"I said we'll be just fine on our own," Prianaj insisted.

"All right then, can't say I didn't offer to help. You two take care now."

Prianaj walked right past the strangers, looking over their heads. Braelyn smiled at Jasira, who nodded back. The strangers hurried past them, rushing to an overgrown bush. Jasira looked back, curious to see what they were doing. Talani must have thought nobody was looking because she lifted a hatch from under the bush, revealing a bunker into which the two disappeared.

Prianaj and Jasira walked along the fence line for about an hour. The fence just kept extending; it seemed endless.

"Maybe those two could have helped us," Jasira said.

"I didn't trust them. Especially the older one," Prianaj replied.

Eventually, they found a gaping hole in the fence big enough for a small animal to wriggle through. Jasira crouched down and prepared to crawl through.

"Wait!" Prianaj shouted.

Jasira crawled back a few steps and sat on the ground. Prianaj picked up a branch that was covered in dry leaves and threw it at the fence. It sparked up and exploded in flames, crackling as it burned. Jasira jumped back and covered her face with her forearms.

"Told you I didn't trust them."

"How do we get through now?" Jasira asked.

"We need to find the generator and turn it off. My guess is those women turned it back on. But the question is, where's the generator?"

"I think I might know," Jasira said.

They started to walk back to where they met the two strangers. This time, they walked faster, shortening the walk by about fifteen minutes. Prianaj recognised the warning signs and the barbed wire at the top of the fence.

"There is no generator here," Prianaj began. "We've been walking forever."

"It's here, trust me," Jasira said.

She looked around the bushes where she last saw the strangers, stomping up and down until she heard a loud clang from under her boots. She dusted away the dirt that covered the large metal hatch with her hands.

"Told you."

"What is it?"

"I saw them go through here. My guess is the generator in here. Do you have another one of those black smoke rocks?" she asked.

"Yeah, I have two more."

"All right then. This is what we're going to do. I'm going to lift the hatch; you throw in the black smoke rocks. Then we hide behind those bushes until the strangers come out. Once they're out, I'll go in and turn off the generator, you keep watch in case they come back. Once its off, we go through the fence."

"Sounds like a plan," Prianaj agreed.

Jasira's feeble arms struggled as she lifted the hatch. Prianaj clenched the black smoke rock in one hand and the basket in the other. The tunnel into the bunker was well lit by a descending column of artificial lights.

"Electricity," Jasira said.

They heard two people muttering, and their shadows moved closer to the entrance. Prianaj threw in one black smoke rock, and it exploded into a huge black cloud. She then threw in the second one for good measure. Jasira shut the hatch, and they hid behind one of the hedges. A few minutes passed, and nobody came out of the hatch.

"Are you sure they're in there?" Prianaj asked.

"Positive. I'm sure of it."

Just as she spoke, the hatch opened and the strangers exited, coughing out black soot. It took them a second to reset themselves.

"That mute girl and the old lady probably got electrocuted," Talani said, coughing up chunks of smoke. "The generator can't handle that surge."

"Which way did they go?" Braelyn asked, brushing away black soot from her face.

"I think that way," Talani said as she pointed in the left direction.

"We didn't go that way," Prianaj whispered.

"Don't worry. It gives us more time," Jasira said, stretching her arm to stop Prianaj from speaking. "I'm going in. Wait for me to come back out."

Jasira crouched down behind the bushes and proceeded to the hatch. The metal ladder was uncomfortably warm.

The majority of the smoke had cleared out, but some of it still lingered. She squinted her eyes so she could see more clearly and tied a cloth around her face to avoid inhaling the smoke. To her right, there were two open doors – one led to a room with two beds, and the other to a room with food and supplies. There was a third door on her left, but it was locked. Jasira tried to manipulate the handle, but it didn't budge. She assumed the generator was behind the locked door.

The bunker had a wide area that had a small dirty couch and stacks of shelves. She rummaged through the shelves to find something she could use to break into the locked room, such as an axe or a bat. She turned heavy crates upside down, sending the contents rattling all over the concrete floor. Most of them were filled with clothes or books, which were of no use to her.

Jasira moved to another pile of crates and noticed that the very top one had some metal bars sticking out of it. She jumped to reach it, but it was too high so she stacked empty crates to climb up. The higher she climbed, the more unstable the stack became. She reached to pull a metal bar but fell, smashing into some of the crates on the ground. Unscathed, she picked up a crowbar and spun it in her hands.

This will work, she thought.

She hit the locked door repeatedly with the crowbar. The wood was almost completely rotten, which made it easier to chip away at it. She knocked off the handle, but another lock kept the door from being opened. Jasira kicked the door and managed to create a gaping hole in it. She continued to kick it until the hole was big enough for her to fit through.

The room was dark, but she felt around the door jamb for a light switch, aided by the blinking lights of a beeping machine. The ceiling lights flickered as they were turned on, revealing the generator in the room. The machine, which almost reached the ceiling, covered the entire back wall.

"Let's turn this thing off," she said to herself.

She fiddled with some of the buttons, but all they did was make the machine noisier. The internal fan spun fast and blew out hot air, which made the bunker hotter than outside. The pipes running above it shook, making clanking noises.

"Prianaj isn't the only one who knows science."

Jasira used to attend general classes in The Border and particularly enjoyed the science classes.

She bashed in the pipes, with the crowbar vibrating in her hands. The pipes cracked, and a warm liquid sprayed out all over her face, with some entering her mouth. It tasted like gasoline, and the strong odour confirmed it. She continued hitting the pipes until the gasoline gushed out all over the generator and entered the crevices. Sparks flew out and ignited the generator. Jasira immediately ran out of the room and climbed up the ladder.

She removed the cloth from her face and breathed in the fresh air. She crawled back to Prianaj, who was still waiting behind the hedges, and warned her to get back down.

"What happened?" Prianaj asked.

"Just wait, you'll see," Jasira replied with a smirk on her face.

Talani and Braelyn were seconds away from the hatch. Too close to not notice.

"They must have gone the other way," Talani said. "What's that smell?" She sniffed the air.

"It smells like it's coming from the bunker," Braelyn pointed to the hatch. "Gasoline."

"Careful, those pipes were old."

Braelyn ignored Talani and entered the bunker.

"Braelyn! You come right back."

"Oh no," Jasira whispered.

"What is it?" Prianaj asked. "What did you do?"

Talani followed Braelyn into the bunker, closing the hatch above their heads. The ground beneath Jasira and Prianaj shook violently, and with a loud pop, the hatch flew hundreds of meters above their heads. It felt like an earthquake. Suddenly, the ground exploded into plumes of black smoke, with the whole atmosphere becoming engulfed with fire and hot shards of metal. Prianaj and Jasira were thrown back by the blast, landing on their backsides. Prianaj caught the basket in her hands and wrapped her arms around it. Pieces of rock and wood landed beside them, missing their heads by inches. The dust settled around them, and Jasira laughed, sitting up and wiping the sweat from her forehead.

"That's what I did," she said.

They sat on the ground for a while, laughing and coughing.

"Somebody would have seen that," Prianaj said.

"You're right. That explosion was a lot bigger than I expected."

"We just wanted to turn off the generator, and you went and blew up the damn thing. You crazy idiot!" Prianaj continued laughing.

It wasn't long before guilt hovered around in Jasira's mind, clouding the thought of if what she had done was the right thing. It was the only option. The more she thought about it, the more her gut wrenched. Evil was never an option Jasira considered, not even once. But she didn't know if it was a monster inside her that woke up, or if it was her conscious protecting her son.

When the afternoon light faded, they made their way to the hole in the fence, running this time instead of walking. Prianaj threw another stick at the fence.

"Just to be sure," she said.

Nothing happened, no flames, no sparks. Jasira entered first, crawling on all fours. Prianaj followed with the basket. They settled for the night underneath a collapsed bridge, and Prianaj tended a small fire. Jasira made sure the ground was soft because she knew it was where they would set up their new lives, away from their past.

PART II

CHAPTER

SEVEN

THE QUEEN'S SISTER'S NECKLACE

SEVEN YEARS PASSED IN THE BLINK OF AN EYE, and so
much of their lives changed. Monday was not like The
Border. It was much larger, and there were more people in the
city. There was no main walkway and no famous underpass
either. However, there were many collapsed overpasses. The
buildings were of the same height and more crowded together,
except for some of the important ones close to the centre of the
city.

Prianaj started another garden just outside the city, similar
to her garden in The Border. The people of Monday bought
her produce using the barter system that had already been

established there. The Border had a similar system, but Monday's was better developed.

They moved into an abandoned apartment block on the outskirts of the city. Jasira and Adam stayed with Prianaj and didn't leave unless it was extremely necessary. Jasira didn't have a job; Prianaj didn't allow it. Having her name on record would have caused major issues, especially if the Cerberi found her after such a long time.

Prianaj's tanned skin had become much more wrinkly, and she had a few more grey hairs than before. Jasira had bleached her hair until it looked like the same blonde hair her sister once had. She covered her face with her hair whenever she went out and sometimes left her face covered even when she got back to the apartment. Jasira rarely cut Adam's hair. She made him leave it long but tied it away from his face and dressed him in generic clothing. He usually wore faded beige cargo pants and a black shirt. His nose was the same as Jasira's – small and slightly turned up.

Jasira headed to the market to buy some new clothes because Adam was growing out of his. He had started to grow faster than normal. Prianaj wasn't home, and like always, Jasira warned Adam not to leave.

"You'll be fine on your own?" she asked.

"Yes," he answered.

The market was about a fifteen-minute walk from the apartment, ten if she jogged, but she didn't like to exercise. Jasira passed a group of Cerberi who walked in the opposite direction. Every time she saw even one, her knees became weak and her palms sweaty. Over time, she got used to them, but she still kept her head down when they passed her. The market was split into two sections – the front for consumable produce and the back for materials. Every vendor was in their fifties or older, and each had their cart set up with their goods either hanging or stacked up high. Jasira recognised Prianaj's cart on the far left side of the market hall. The enclosure was a thin aluminium shed with cut-outs for ventilation and light – three large square holes on each of the four walls.

"What are you doing out of the apartment?" Prianaj asked in a breathy voice.

"I need more clothes for Adam."

"Shush."

"I mean, for my daughter," Jasira corrected.

"Be quick, you know how restless children are if they're alone for too long."

"I will be, I know what to get. Just shirts and pants. His boots should be fine for another year."

Prianaj nodded and served a woman who approached the cart. Jasira ducked under the fabric that conjoined two carts and moved to the back of the market where the clothing stalls were.

She stopped at a stall with wool jumpers and blankets that were stacked in a small pile. The winter months were about to roll in, so she bartered with bread rolls and biscuits in exchange for winter clothing. A few stalls away, she saw piles of unfolded shirts of all different shapes and sizes. Jasira picked out a few oversized ones that Adam could grow into over the next few months.

"Take them, free of charge," the saleswoman said. She had black hair and silver eyes, the same silver eyes as Prianaj. She was the same height as Jasira and had thin arms and legs.

"I couldn't possibly. It must have cost you a fair bit to source them," Jasira said, handing the shirts to her.

"They're of no use to me now. Please, I insist. I'm planning on leaving in a few days. Bartering with them won't get me much," she said, handing Jasira the shirts back.

"Why do you need to leave? Monday is the best city in Diektra."

"Seven years ago, I left my home in search of something, rather someone," she began as she packed up her goods. "But they're not in this city so I have to try The Border."

"Didn't they tell you where they'd be?"

"I'm not sure I know who I'm looking for exactly, let alone where they could possibly be. But that's too much information I've already told you."

"Thank you for the clothes. Please, come with me. I can give you some fruit for your travels," Jasira insisted.

The woman followed Jasira through the labyrinth of stalls to Prianaj's cart.

"Prianaj, this lovely lady has just given us a heap of clothing. Can we give her some fruit?"

Prianaj packed a mix of apples, oranges and berries in a small basket and thanked the kind woman. The woman leaned over to grab the basket and her necklace fell out from underneath her shirt and hung in the air. Before Prianaj could hand the basket to her, she dropped it and the fruits smashed onto the ground, splattering into a shallow puddle of juices.

"Your necklace," Prianaj said as she reached to touch it with her finger.

Jasira kicked away some of the fruit that landed on the other side of the cart. The woman wore a gold necklace with a red gem in the centre.

"Where did you get that?" Prianaj asked.

"It's... a family heirloom," the woman answered, tucking it beneath her shirt.

"I had one just like it. That means yo—"

"Impossible. There are only three in existence, all of which should belong to people in my family. I'm sorry, but I have to go," the woman said, rushing off to the back of the market hall.

"Jasira, get up," Prianaj began as she moved the cart to the side. "That woman knows something, I can feel it."

Prianaj and Jasira scurried to the woman's stall, almost slipping over the splattered fruit. To their surprise, however, it was abandoned, and all the clothes had been removed.

Prianaj never spent more than five hours at the market. Customers only bartered in the morning, and she couldn't stand for too long. Even though there was a little wooden stool she sat on during quiet times, her bones were not as strong as they used to be.

Jasira pushed the cart round to the back of the apartment building, leaving it out of view from any passers-by. Adam finished eating a bowl of stew Prianaj had prepared earlier with vegetables from her garden.

"Can he not go outside for even half an hour?" Jasira asked. "He's a kid and should enjoy his childhood and not just be stuck in here all his life."

"Only if you want him to be caught by someone, or worse, killed by the Cerberi," Prianaj warned. "You wouldn't want people finding out about our secret." She hushed her voice, fearing somebody might hear them through the thin walls.

"Secret? If you want to start talking about secrets, I think you should go first."

"I have nothing to hide," Prianaj said.

"Adam, please go to your room," Jasira said firmly as she and Prianaj moved to the kitchen area. "First, you didn't want to leave the Border seven years ago. Now we've come across a woman who has a necklace that you apparently had once upon a time. I mean, your face when you saw it, you were shocked."

"Just maybe, Jasira, there are things in my life that I don't want to share."

"Then why stay? You could have left the moment we arrived here," Jasira said.

"You wouldn't have made it half as far as you did without me. You would have gotten the boy killed. I only stayed because I made a promise to your sister."

There was a knock at the door, and Prianaj froze in place.

"This conversation isn't over," Jasira said as she tiptoed to the door.

Jasira looked confused as the woman from the market invited herself in.

"Alright then, make yourself at home," Jasira said sarcastically.

"I'm sorry, I followed you both here. I need to speak to that other woman."

"You can't just walk into a stranger's home," Prianaj said, moving towards the single armchair in the main room.

Jasira pulled up a stool for the woman as she sat on the ground.

"Actually, I can," the woman said, taking off her headscarf. "My name is Elaine Weiluk."

"Royalty?" Jasira asked with wide eyes.

"The queen's sister," Elaine answered.

"The queen doesn't have a sister," Prianaj said quickly.

"My mother disobeyed the law and gave birth to a second child, me. She was killed for her disobedience. But I presumed you knew that, since you recognised my necklace. Unless, no, you were banished way before my birth."

"Banished? Banished for what?" Jasira asked.

"For being born," she responded.

"Wait, were you in the citadel?"

Prianaj didn't respond.

"Couldn't your mother have changed the law, to avoid regicide?" Jasira asked Elaine.

"My mother wasn't queen at the time, and a unanimous vote was required, so she raised me in secret," Elaine began. She spoke a lot with hand gestures. "My grandmother ruled these two cities. She too had a sister but banished her for fear of losing power. After her own daughter disobeyed the law, my grandmother feared people would see it as weakness if she showed mercy, so they had to punish her... with execution. My mother was then also killed for breaking the same law, and my sister Samaari claimed the throne for herself."

"Why are you here, then?"

"I saw how dangerous the citadel can be, so I ran as far as possible from it. I set up shop here for seven years and haven't left ever since."

"You barely know us. Why are you telling us this?" Prianaj questioned.

"If you really had the necklace, I must trust you. It's one of the laws of the holder."

"What if I lied, about having the necklace?" Prianaj said as she raised her eyebrows.

"Then you would have broken the law. But you already knew that. Tell me, how did you come into possession of one of the necklaces? Did you take it from my dead grandmother? Because my mother would have had one, but my grandmother never passed it down."

"Why are these necklaces so important anyway?" Jasira asked.

"They were made—"

"They were forged before the end of the Culling," Prianaj interjected. "Three were created each from a different metal; gold, silver and bronze that was melted from the dying embers of a burning fire. One for each of the three women of royalty at a time. They were to be passed down from grandmother to grandchild before their death. It was a symbol for each of us to trust each other. But that was destroyed when we all broke the law."

"Us?" Jasira asked.

"Us," she repeated. "I am Prianaj Weiluk, sister of Marya Weiluk and daughter of Estella Weiluk. Marya gave me her necklace," she announced.

"You didn't tell me you were royalty," Jasira said. "Talk about secrets."

"Leader of the Cerberi, your position before your exile. Why didn't you ever come back?" Elaine asked.

"I was exiled, I could never return."

"Yes, but Marya is no longer queen."

"I suppose Samaari will let me stay, then?"

"No, she probably won't. She's as wicked as all the rulers before her."

"Why did you come here, then?" Jasira asked.

"I told you, I had to get away from the citadel."

"No, I mean, here. To find us," Jasira repeated.

"Seven years ago, when I left, my sister was told that the prophecy of the male child had been fulfilled, but she didn't believe it to be true."

Jasira glanced at Prianaj, knowing that Adam was only in the next room.

"I also left for another reason," Elaine continued. "I wanted to see if the rumours were true. I wanted to find the boy. If he had been born, then perhaps he would be the living proof that the world is healing."

"And you don't want the boy dead?" Prianaj asked.

"No, of course not. I could have protected the boy under my authority. No harm would have come to him. For years, I was taught that the Culling was nature's way of ridding the earth of man because of their terrible ways. But for those same years, I realised that the women who preached that had become drunk with their own power."

"Hypothetically, if I knew where the boy was, if I knew him at all, what would you do?"

"Jasira, no," Prianaj said.

Elaine took off her necklace and wrapped it around her hands, with the pendant hanging. She placed her hands in Jasira's.

"I swear that no harm will come to the boy from my hands. He'll be under my protection, he'll be under the protection of the crown," Elaine promised.

Jasira ambled to Adam's room and opened the creaky door.

"Adam," she called.

He walked out, holding onto his mother's hand.

"Elaine. This here is Adam."

Elaine jumped up, her face flushed with joy. She danced a little and screamed in excitement.

"Shush!" Prianaj and Jasira said in unison.

"Adam. What a gorgeous name."

"Who's that?" Adam asked as he hid behind his mother and used her skirt to cover his face.

"This is my niece, Elaine," Prianaj replied, knowing it was the first time in a long time anybody used that word.

"My God. He's real. Your grandson?"

"No, he's not my blood," Prianaj said as she shook her head.

"Oh, sorry. How did you find him?" Elaine scratched her head.

"He's mine," Jasira said.

"Wait a second. So, you're not Prianaj's daughter?" she asked Jasira.

"No, Prianaj is not my blood."

"But if Jasira has no royal blood, then the boy would not rightfully inherit the crown. Prianaj, you would have been pregnant, you had the Procedure before you were banished. Is she really not your daughter?"

"My child died at birth. Jasira can't be my daughter because her eyes are blue, and mine are silver. My child would have the same eyes as me," Prianaj said.

"And Adam has blue eyes as well, so he can't be of royal blood," Jasira said. "Wait, Prianaj, I thought your child was stolen, not dead."

"My baby was stolen, then killed." Prianaj said dismissively.

"But the prophecy was fulfilled anyway. A boy was born in March during a blood moon," Elaine said.

"Eye colour might not matter," Elaine said as she began pacing around the room. "Look at Prianaj and myself. I am much more pale than her."

"Skin colour doesn't carry over all the time during the Procedure, but always eye colour," Prianaj responded.

"Not of royal blood, though," Jasira responded as she took Adam back to his room.

"This is a lot to process at once. My aunt is alive and well, with the boy I have been looking for for over seven years. But maybe he isn't the right boy... This doesn't sound right," she said, shaking her head.

"Perhaps, take some time to process all this news, then come back," Prianaj recommended.

"No, let her stay here. She's family," Jasira insisted.

"My family shut me out a long time ago. I would have died for them a long time ago. But now they are dead to me," Prianaj said firmly.

"Then don't make the same mistakes they did."

CHAPTER

EIGHT

ABDUCTED BY SPIRITS OF THE PAST

THE APARTMENT REMAINED DARK for much of the day,
even when the morning light shone through the small
windows. The only lights that illuminated the rooms at night
were the candles around the apartment. It didn't take long for
Jasira and Prianaj to get used to the inadequate lighting; they
were already used to it from the cave. Adam had been in the
apartment his entire life so he had nothing to compare it to.

Elaine rolled over in a blanket in the main room, using a
cushion from the armchair as a pillow. She covered herself with
one of the blankets Jasira traded the previous day.

Jasira was the first to wake, tripping over clothes on the ground. She noticed Elaine folding the blanket.

"You're up early," Elaine said.

"I'm always up this early," Jasira replied.

"I just sleep until I feel well rested and then I get up."

"If only we all had that luxury. It gives me peace of mind to see Adam is still here," Jasira said, pouring warm water into a metal cup.

"Is that coffee? I could do with some."

"I wish. But no, we ran out of coffee a while ago. I have tea, though. Prianaj grows green tea leaves in the garden."

"Tea would be great."

"So when are you planning on leaving?" Jasira asked, handing a mug to Elaine.

"Pardon me?"

"At the market, you said you were planning on leaving in a few days."

"Oh, yes. I came here to find the boy, but after seven years of bad luck, I thought there was no hope. I was planning on moving to The Border if I couldn't find him here."

"Prianaj and I came from The Border," Jasira began. "We moved here after the Cerberi raided the city."

"They searched here too but gave up after a while."

"So you are the queen's sister? What's she like?" Jasira asked.

"Terrible, controlling, self-centred, all the bad stuff." Elaine laughed. "I doubt she's changed over the years. I blame the curse."

"Curse?"

"It's a rather long story, more of an inside joke," Elaine said.

"I had a sister once, an adopted sister," Jasira began.

"Much better than Samaari I can only assume."

"She was kind and loving. Her mother took me in when I was a baby," Jasira continued as she sipped tea. "But then her mother passed away so it was just me and Vaika for as long as I can remember. She took care of me when no one else would."

"Your sister's mother must have raised her well."

"Vaika described Sirena as a good mother. I trust her judgement."

"Sirena?" Elaine asked as she pulled the cup away from her mouth and raised her eyebrows.

Adam came out of his room, stretched his arms and yawned loudly.

"Hey buddy," Elaine said, rubbing her hands in his messy hair.

Jasira combed his hair with her fingers and tied it up into a little bun. Prianaj also came out of her room, rubbing her eyes with the back of her hand. The room was quiet for a while as they moved around in silence. Elaine handed her empty mug to Jasira, almost laughing to ease the tension in the room.

The only noise was the sound of the water Prianaj poured into her mug.

"Well, I better get back to my home," Elaine said, breaking the silence.

"How do I know you won't tell anyone about the boy?" Prianaj asked.

"There it is." Jasira snickered.

"I made a promise, with the necklace," she replied, taking out her necklace from underneath her shirt. "He's protected."

"Prianaj, if Elaine says that Adam will be safe, then we should trust her," Jasira said.

"I don't. The boy stays, and you will too," Prianaj said.

"Somebody woke up on the wrong side of the bed."

"Jasira, please not now."

"No, it's fine," Elaine said. "She has no reason to trust me. I'm a complete stranger to you both."

"The boy needs to be kept safe. I don't want Cerberi barging down the door to take him away," Prianaj said.

"Prianaj, stop it. You're being disrespectful to our guest," Jasira began. "She promised she wouldn't tell anyone about Adam, just leave it at that."

"I don't need help defending myself, Jasira."

The apartment shook with a loud bang as the front door slammed shut.

"What was that?" Jasira asked.

"Adam, go to your room!" Prianaj commanded.

"Adam!" Jasira repeated.

Elaine turned to check the armchair where he sat but only found the wooden figurines he played with. They ran out to the corridor, but it also was empty.

"You scared him, Prianaj," Jasira said as her blood rushed to her face.

"Just calm down for a second. Is there anywhere he likes to go?" Elaine asked.

"He's never been outside. Now someone will find him."

They sprinted down the uneven staircase to the ground floor of the building missing few steps as the railing came loose. At the intersection directly opposite the building, Elaine went left, around the back end of the apartment block, and Prianaj and Jasira went right, towards the main city. People flooded the streets from every direction.

"I swear, Prianaj, if anything happens to him—"

"We will find him."

They weaved through a row of women pushing carts to the market hall, looking over their heads to scan the area. Prianaj struggled to keep up with Jasira's pace, stopping every few seconds to catch her breath. They pushed past groups of women in the middle of the path, accidentally bumping into teenagers fixing their hair in the mirrored windows of the bank building. People rarely entered that building; only the rich used the banking systems. The lower class stuck to their bartering and trading in the markets.

A woman who looked the same age as Prianaj blocked their path.

"Prianaj, aren't you headed for the markets?" she asked.

"Not today unfortunately," Prianaj replied.

"We're a bit busy right now. Excuse us," Jasira said as she forced the woman to the side and continued to search for Adam.

A group of Cerberi noticed them scrambling through the crowd and stopped them, hitting their batons against their palms.

"You two seem to be in a rush," one of the Cerberi said.

"Just lost my... my daughter."

"We must help you search," the other Cerberi said as she looked over people's heads.

"That won't be necessary, she always does this. Oh look, I think I see her," Jasira lied, hurrying past them and running sideways between people.

Jasira jumped up using Prianaj's shoulder to push herself. She noticed a gap in the crowd, and the women gave bizarre looks as they moved along. She dragged Prianaj by her wrist to where Adam was sitting with his hands over his eyes. Jasira knelt down and rubbed his shoulder. His long hair confused the bystanders. Some would have thought Adam was a girl, but from the angle Jasira looked at him, it became obvious he was a boy. The women whispered to each other with bulging eyes and gasps that grew louder. Most of them covered their own shocked faces with their palms, but the ones that stared at him for too long frightened Jasira the most.

"Hey," she said. "It's all right, mama's here."

"Oh, thank God you found the child," Prianaj said, leaning on her walking stick.

Adam held his mother's hand as they walked back to the apartment, taking a route around the perimeter of the city rather than through the crowd.

"No one is like me," Adam mumbled.

"What was that?" Jasira asked.

"No one is like me," he repeated.

"What do you mean, honey?"

He stood on the spot, looked down at the pavement and kicked his boots against the ground.

"What does he mean by that?" she asked Prianaj.

"He just got lost in a crowd and didn't see any men. I would be very suspicious if I was the one."

They hurried around the back of the apartment, turning at the sound of Elaine's voice.

"What a blessing. You found him," she called out.

Prianaj reached for a handrail but was thrown back into a tall pile of garbage by a blast from the back entrance.

Elaine screamed and ran toward them, but a second, much larger explosion thrust Jasira and Adam to the ground and knocked them unconscious.

Jasira struggled desperately to breathe. Icy-cold water was flooding over her. Her throat burned as if she had swallowed hot coals. Her body swirled around as she cried for help, but only bubbles of air came out of her mouth. She opened her eyes, but everything was a blur. Deep, red water swallowed her whole. The dream she hadn't dreamt in seven years felt real, again.

Jasira's head was pounding, and everything was a blur. She was in the centre of a tent, and her wrists were tied around a wooden post with a rough rope that dug into her skin. Prianaj was tied on the other side of the post with her back against Jasira's. She noticed Adam on the other end of the tent, asleep with his hands tied to his feet and a cloth that covered his mouth.

The tent cloth was a creme tint that filtered the sunlight into a deep amber. For a second, Jasira thought she was back in the cave. There was no ventilation, except for a small opening at the front of the tent. Other than that, they were just recycling the air they exhaled.

"Prianaj, are you awake?" she whispered.

"Alive and well. What about Adam?"

"He's tied up in front of me. What happened?"

"Looks like we've been abducted."

"Abducted? By who?"

"My guess is the one person who's not with us," Prianaj said, wriggling to try and free herself.

"No. Elaine was with us when the building exploded."

"But she's not here is she?"

"You never gave her a chance, Prianaj. She finally got some of her family back, and you did too, but you pretended she didn't exist," Jasira said softly.

"Arguing won't do us any good in here," Prianaj said, still trying to wriggle herself free.

"I can't get my hands free, can you?" Jasira asked.

"The rope is too tight."

"Hang on, I think I see a knife."

Wedged into the wooden post to the side was a small silver dagger. Jasira stretched her legs, which surprisingly were not tied, and kicked the post. The dagger wobbled but remained attached to the post.

"Did you get it?" Prianaj asked.

"No. Do you see your walking stick? I need something to push it out."

"No, but this might help," Prianaj answered, pushing a piece of wood over to Jasira using her foot.

"Yes, this might do," Jasira said.

She kicked it up against the post right under the dagger. With one hard strike, she kicked the piece of wood and sent the dagger flying through the air. Prianaj and Jasira ducked their heads as the dagger bounced off the post they were tied to and landed next to them. Jasira couldn't reach it with her legs and neither could Prianaj because it had landed directly next to them. Dust was kicked into the tent, and a hand reached for the tent flap.

"Someone's coming," Jasira shrieked.

She quickly leaned over, pressed her lips on the dagger and grabbed it with her teeth, flicking it behind her into Prianaj's hands. The woman that entered wore a white veil that covered her face. It was connected to the rest of her silk dress, which was the same colour. Her feet were bare and missing a few toes.

"Who are you?" Jasira asked, bending her knees towards herself.

The woman didn't reply; instead, she reached for Adam.

"Don't touch my child!" Jasira screamed.

The woman looked up. Although her face was covered, she emanated a palpable sense of authority.

"What do you want with us?" Prianaj asked.

The woman bent over towards Jasira and breathed into her ear. Her breath was cold and loud. Jasira lifted her shoulder to stop the air from touching her skin, but the woman pushed her shoulder back down. Jasira felt her own pulse in her fingers, and her breath became shallow.

"I guess cat ain't got your tongue," the woman said, struggling with the words. Her voice cracked as she forced them out. "Years ago, you two took something from me," she croaked, pausing after every few words. The woman's voice sounded familiar to the both of them, but they couldn't quite put a finger on it.

"Still think it was Elaine who abducted us?" Jasira asked Prianaj.

"I haven't ruled that out yet," Prianaj said.

"You must be mistaken, we rarely meet people," Jasira said.

"Now I will take something from you," the woman continued as she pulled a dagger from her thin belt and held it to Adam's neck.

Jasira felt Prianaj struggling to cut through the rope because the blade was blunt.

"No!" Jasira screamed. "Don't hurt him."

"Him?" The woman asked, staring at Jasira through her veil.

Jasira didn't reply.

The woman looked at Jasira, then at Adam, and then back at Jasira. She separated the hair from his face with the tip of the dagger and moved it behind his ear.

"Oh my, the rumours are true, and you didn't tell me seven years ago," she began as she tucked the dagger back into her belt. "I have a better idea. I'll just turn him in to the Cerberi. They'll want to meet him."

"No, please. What do you want? Food? Money?"

"My daughter," she replied.

"Huh?"

The woman removed her veil, revealing her severely burnt skin and a scar that stretched from one ear to the opposite cheek. Her face had melted onto itself, covering her nostrils and blinding her left eye. It had dried into scabs like dry leaves waiting to crack. She had no hair on her head, but Jasira recognised her even though her face was badly disfigured.

"Oh my God!" she gasped. "Talani."

"Talani?" Prianaj repeated, still attempting to cut the rope. "She died in the explosion."

"Burning shrapnel sliced my body. I wore bandages for three whole years, not knowing if my daughter even survived." Talani's voice grew louder, and her tone became harsher. "It wasn't until my rescuer told me that my daughter's blood was smeared across the plain. I lost everything that day. Do you even know what it's like to lose a child?"

"I'm so sorry, Talani, but I wouldn't have had to do that if you two didn't want us dead. It wasn't just a coincidence the electric fence turned on."

"Out in the wild, it's kill or be killed. You two were just unlucky to walk into our path."

"What are you going to do with us?" Prianaj asked, quivering.

"I'm going to turn in that... that boy." Talani began. "But first, I'm going to slice your face, the same way you sliced mine."

Talani pulled out her dagger and pressed it against Jasira's cheek. She pierced Jasira's skin, and a stream of blood trickled out.

"I might cut the boy's face too. So he can feel the pain I felt."

Tears began to fall from Jasira's eyes, mixing with the blood as her body shook. Talani wasn't moving the blade, but Jasira's slight movements caused the blade to slice her skin. She closed her eyes, and the pain stopped. Suddenly, glass shattered all over Jasira, and some pieces fell into Prianaj's hair.

"What was that?" Prianaj asked, finally breaking free from the ropes.

Jasira opened her eyes and saw Elaine's bleeding hands untying the ropes around Adam's legs.

"Get away from him, Elaine!" Prianaj demanded.

She lifted her hand, which held a dagger, and Elaine raised her hands. Prianaj cut the ropes around Jasira's wrists.

"She's here to rescue us," Jasira said, rubbing her wrists.

"She's not here to rescue us," Prianaj retorted. "She's with Talani."

"I don't even know that woman," Elaine said in defence.

"Liar," Prianaj said.

"I swear," Elaine pleaded, holding her necklace.

"How did you find us?" Jasira asked.

"After the explosion, I was half conscious and saw this woman dragging you three into a cart and taking you out into the desert. I followed her here," she answered, pointing to Talani, who lay flat on the ground.

Over the years after the Culling, the earth dried up, and most of the land became desert. The mountains thrived with plant and animal life because they were closer to the heavens.

Jasira screeched for Elaine to move out of the way as Talani swung her dagger towards her. The dagger dug deep into Elaine's thigh and poked out the other end as it slashed out a huge chunk of flesh. Prianaj threw her dagger right past Jasira's ear and into Talani's breastbone, killing her on impact. Blood bubbled out of Elaine's leg and gushed onto the ground. Jasira rushed to her side, slipping in the pool of blood, and tried to comfort her.

"Elaine, stay with me. We need to get help!" she yelled at Prianaj.

"No, Jasira," Elaine began.

"But you'll die, you're losing too much blood."

"Listen to me," she continued. "Take my necklace, it will protect the boy."

"Elaine, no, I'm not royalty. I can't."

"Take it. If the Cerberi ever catch you or Adam, show them the necklace. They'll have to take you to the citadel. He's still protected under my authority. Nobody will be able to take the necklace from you unless you take it off yourself."

"We can get you help, the Cerberi can take you back to the citadel."

"No. I can't go back. I haven't been there in almost a decade. Even if I do go back, Samaari will want me dead for disobeying her orders," Elaine said as she tugged the necklace from her own neck and handed it to Jasira. She looked at Prianaj, who was untying Adam.

Prianaj stared at Elaine with forgiving eyes that tried to accompany a half-smile, but she turned away quickly.

"It was nice to finally meet you, Prianaj, thank you for letting me stay," she said as her voice faded out.

Jasira held Elaine's blood-covered arm as her strength also faded. She felt Elaine's heartbeat slowing through her wrist.

"Jasira. The boy is proof."

"Proof of what?"

"He is proof—," she whispered, taking her last breath.

Elaine's eyes remained open as her body lost its strength. Life slowly faded from her eyes. Jasira let out a soft cry and gently closed Elaine's eyes with her hand.

CHAPTER

NINE

PRACTICE MAKE PERMANENT

P RIANAJ COVERED ELAINE'S GRAVE with the last bit of
earth they dug up. Jasira placed a bouquet of dahlias on
the grave and stood for a few moments in silence. They were the
same flowers Jasira picked out for Vaika's burial – the burial she
never attended.

"I'm going back to The Border," Jasira said.

"Why? There's nothing there for us, Jasira."

"I have to see my sister's grave. I'll only be away a few days,
so you'll have to watch Adam," she said as she put the necklace
around her neck.

"You're going alone?"

"Vaika had something, a box she never let me open. She always hid it from me. I think it's important. Maybe there's something inside it about my past, where I came from, who my mother is."

"Someone might see you. It's too dangerous," Prianaj warned, leaning on her walking stick.

"It's been seven years. Nobody will recognise me. I'll be in and out."

"What if the contents of the box are meaningless, what if the box is not there at all?"

"It's got to mean something, or she wouldn't have hidden it from me. If the box isn't there, then I still get to see my sister's grave."

"I'll come with you then. I won't let you open pandora's box all by yourself. Who knows what you'll find."

"And who will look after Adam?"

"He can come with us. I'm sure we can hide him like we did the first time," Prianaj suggested.

"He was just a baby then. It will be impossible, and it will be too suspicious for all three of us to walk in and out of The Border."

"Where there's a will, there's a way."

"No. This is something I have to do alone," Jasira said.

"Jasira Revken, you listen to me right now," Prianaj growled, slamming her foot down. "You're weak and cannot defend yourself if you're in danger," she continued, this time hitting Jasira with her walking stick.

"Ouch. What was that for?"

"Weak and defenceless."

Jasira charged at Prianaj and raised her hand to strike, but Prianaj turned quickly and pushed her out of the way.

"Fine then. Come with me."

"No. You will learn to defend yourself."

"And you're going to teach me?" Jasira asked, regaining her balance.

"I led the Cerberi for a long time in my early years. There isn't anyone better to train you. We'll get up at the crack of dawn every morning, and perhaps given a week's training, you'll be fit to travel on your own."

"I'm already used to waking up early. It won't be hard."

"Of course not. The early mornings are easy. It's the physical training you need to get used to."

"When do we start?"

"Now!" Prianaj said firmly.

Prianaj was usually the first to get to the market shed of all the other vendors, so she knew the best times to train there – early in the morning and later in the afternoon when it was empty. Jasira took a striking commitment to Prianaj's training, waking up before sunrise to jog around the city, and making it to the market shed early.

"Do you really think I'll be ready in a week, Prianaj?" Jasira asked, opening the large metal door to the shed.

"That's up to you. Whatever amount of effort you put into the training won't matter as long as you believe you can do it. Practice, practice, practice."

"Practice makes perfect."

"Wrong. Practice the wrong things over and over, and there is no perfect. Practice makes permanent," Prianaj said, throwing a dagger into a wooden column they used as a makeshift target.

"How did you make that shot? I can barely see that."

"Train in the dark, and your eyes will adjust."

The market shed had no real lighting. Even during the day, when a few beams of sunlight streamed through holes in the walls, the shed remained quite dark.

Prianaj handed Jasira a set of three daggers. Jasira stood in place for a moment, unsure what to do.

"Go," Prianaj motioned for Jasira to throw them at the target.

"Right, uh, okay."

Jasira stretched her left arm out in front of her and pointed at the target. She squinted her eyes to get a better look at it, but it was still too dark. She threw the dagger at the target with her right hand, and it disappeared behind the wooden pole.

Prianaj rolled her eyes and scoffed. "Again!"

Jasira tried a second time; this one scratched the side of the pole but also disappeared into the shadows. Prianaj exhaled loudly as if she was bored with Jasira's futile attempts.

Jasira picked up the third dagger, this time closing one eye and straightening out her shoulder blades. She let the dagger fly from her hand, cutting through the air and lodging itself in the very edge of the target.

"Again!" Prianaj commanded, handing Jasira another set of daggers.

"That was good, wasn't it?"

"Unless you can hit the target with your eyes closed, it was terrible."

Prianaj stood next to Jasira and turned her face to the side. She threw a dagger at the target without looking and scored a bullseye. Jasira stood in awe of her.

"Again!" Prianaj repeated.

Jasira threw dagger after dagger, and for each miss, Prianaj made her run a lap across the shed. Jasira was out of shape, and exercise was not one of her strengths.

At the end of each training session, Jasira was covered in sweat and had lost a fair bit of weight.

By day four, Jasira was used to the training, and it became second nature to her. She was able to hit the target fifty percent of the time, and if lucky, she hit the bullseye, but that was extremely rare.

"How long were you the leader of the Cerberi?" Jasira asked as she yanked out daggers in the wooden pole.

"Nearly two years in my late teens. And I was one of the best."

"How many did you have under you?"

"Hundreds. Trained them all at the citadel for a few months before they got allocated to a city."

"Where's the citadel?"

"Depends on who's asking. But that's a secret I can't reveal. I was sworn to secrecy," Prianaj said as she clasped her fists around her walking stick.

Jasira dropped all the daggers in her hands and blocked her head from being struck. Prianaj kicked the sides of Jasira's legs and then hit away the daggers. Jasira crawled away and covered her face from the dust that was blown up, stretching her arm out to stop Prianaj.

"Your defence is slow," Prianaj said, trying to catch her breath.

"That's going to leave marks all over my body. I'm probably going to need a healer after all this."

"Real fighters don't need healers. Defence is more important than offence, but we started with knife throwing because that's what I was best at."

Jasira was a fast learner. Even from a young age, she was determined to learn whatever she could quickly. It was a skill Vaika taught her.

By day six, Jasira was able to sense attacks from all directions. Her muscles ached all over, but she reminded herself of what Prianaj had said to her earlier in the week, "whatever amount of effort you put into the training won't matter as long as you believe you can do it." Jasira pushed through her pain and pretended it didn't exist. She made sure she practised her drills properly, as Prianaj warned her earlier that incorrect practice would lead to imperfection.

Jasira's offence improved significantly within a few days of training, and she was able to hit the target one hundred per cent of the time. But her training was far from complete, even though there was only one day left.

It was day seven, the final day of training, and if Jasira managed to accomplish her final task, then Prianaj would allow her to go back to The Border.

"One more test, Jasira," Prianaj said as she tied a cloth around Jasira's eyes. "I said when we first started this training that you wouldn't be ready until you could hit the target with your eyes closed. No cheating."

"The shed is dark enough," Jasira said.

"Your eyes would have adjusted to it."

"How did you come up with this theory of knife throwing without vision?" she asked.

"My trainer was blind. Even without her sight, she could somehow sense where targets were and she never missed a shot. She liked to think it was her superpower. I guess after the Culling, anything would have been possible. I learned like her, blind, and I've never missed a shot either," Prianaj responded.

Jasira felt the warm metal daggers in her palms and exhaled slowly as the silence grew louder around her. She could hear the crunch of dirt beneath her boots, and she tilted her head up. She held her palm out and tried to sense where the target was. Her hand tingled, and she instantly knew she was right in the direction of the target. She held one dagger to the side of her ear, breathed in quickly and then exhaled slowly, very slowly. The air beside her whooshed as the dagger left her fingertips. She heard it land with a muted thud into the target.

"Two more," Prianaj whispered.

Jasira turned and walked away from the target. She twisted her torso around and then let the second dagger fly. She heard it land with another thud into the target.

"One more," Prianaj said. "Let's see how well you do with obstacles."

Jasira heard Prianaj's feet shuffle the dirt behind her. The air whooshed as Prianaj swung her walking stick at Jasira's head. Instinctively, Jasira ducked in time and rolled onto the floor, landing with her outstretched arm holding the last dagger in front of her.

With each step Prianaj took, Jasira mirrored her. Round and round in circles they went until Prianaj attempted to strike with her walking stick. Jasira managed to block each strike. She heard cracking sounds and hoped that was Prianaj's stick breaking apart rather than her own bones. She'd never experienced any bone fractures and was determined to keep it that way.

Jasira threw the last dagger and heard it lodge into the wooden pole.

"Three," she said, huffing and puffing before finally catching her breath.

Jasira took off the blindfold and looked at the target in confusion. There were only two daggers wedged into it. Prianaj appeared in front of her with her walking stick in both hands.

"You are ready to go alone," she said, pulling the third dagger out of the head of her walking stick.

Samaari threw her cup across the mahogany dining table and stabbed the wood with her knife. A Cerberi walked into the room uninvited and swung open the doors to the dining hall.

"Why do you interrupt my meal, Orbina? This better be news about my sister," Samaari said in frustration.

"Apologies, my queen," Orbina said confidently. "But unfortunately not. Idira is still with the others searching for her."

"Then why are you here?" Samaari asked, trying to eat, although she never could if she was being watched.

"I bring news from The Border."

Samaari lifted her head and pointed her ear in Orbina's direction. Her ear twitched, and her lips parted slightly as she rested her elbows on the table and her knuckles beneath her chin.

"Speak," she said.

"We have found a box."

"A box." Samaari said, laughing.

"With the royal engraving," Orbina continued. "It was found in a building at the edge of the city."

"Show me this box, I want to see it."

"My queen, we have it safe in The Border. We can't be sure of its contents."

Samaari rose slowly from her chair, her sharp nose twitched once.

"You will go back to The Border, and you will bring me this box."

"Yes, my queen," Orbina responded, quivering. "But it might be a while, the masquerade is in a few days and the box has the royal engraving, so it will be the centrepiece of the event."

Samaari blew out three candles, one at a time, on the table as she walked across the room. She lifted a fourth and held it in front of Orbina.

"Bring me the box," she demanded, before blowing the candle smoke into Orbina's face.

CHAPTER

TEN

THE BOX

T HE BACKPACK THAT JASIRA HAD USED for the last
decade faded from its original dark black colour to sandy
green. The backpack had also started to grow mould from all its
travelling over the years. Jasira was packing and preparing as she
usually did for her long journeys, but this would be her first
solo trip ever.

Prianaj diced fruits and vegetables and sealed them in metal
containers. Surprisingly, everything fit inside the backpack,
including a spare blanket in case of cold weather. It rarely got
cold, though. During the summers, the weather was generally
warm throughout the day and even at night.

Adam ran out of his room and threw his arms around his mother's waist.

"Aw, I'll be back in a few days, buddy," she said reassuringly as she combed her fingers through his hair.

Prianaj cut his hair shoulder length to stop it from getting in his face all the time. There was no point anymore in keeping it too long; he rarely left the apartment.

It was the first time Adam was going to be without his mother for more than a few hours. Jasira was unsure if Prianaj was the best person to take care of him while she was away, but there was no other choice. If Elaine hadn't died, she would have left him with her. Jasira felt she would have been able to trust Elaine more than Prianaj, even though she had only known Elaine for just one day and Prianaj for nearly a decade. Perhaps it was because Elaine was so honest and open about her opinions.

Luckily, the side of the building that exploded did not destroy Prianaj's apartment. Adam still had a safe place to sleep at night. Despite the explosion and abduction, Jasira believed it was safe for Adam to spend the next few nights there without her.

Jasira covered the top of her head with a thin scarf, leaving some strands of her hair loose to cover her face. Prianaj munched a red apple and wished Jasira a safe journey. The door closed softly, clicking as it locked.

Jasira stopped at the electric fence, which was now almost entirely rusted, with most of the panels overgrown with vines. A part of the fence had fallen over and created a direct path to The Border. The bunker from the explosion years ago was empty except for the tiny scraps of metal inside. Jasira stood on the edge and stared into the crater. Then she looked up and saw the old hatch some meters away. It was bent and severely corroded. She dragged it to the bunker, balanced it on the edge and pushed it over with her foot. Rocks and sand slid into the bunker along with the hatch, rolling slowly down to the bottom. Jasira finally felt the weight of the strangers' deaths lift from her shoulders.

She made it to the city entrance, which was still unguarded. It remained freely open so that citizens could gather resources or water from the mountains. The main walkway had been renovated with fresh, dark-grey concrete. People's carts no longer got stuck in the potholes, and there were no more cracks for anyone to trip over. There was a tall scaffold attached to the city's central tower, which Jasira had never seen before. Workers were attaching glass panels to all four sides of the tower, from the top to the bottom.

The tunnel to Jasira's old home was blocked off with timber panels that were criss-crossed at the entrance. The wooden back door had been replaced with a steel door and was also barricaded with vertical steel posts. The window beside the door, which belonged to Jasira's old room, hadn't been blocked off. It had stained yellow over time, and the corners had darkened with dirt and dead bugs.

Jasira reached for a hefty piece of rubble from the alleyway, which was never cleaned by anyone, and with plenty of force, she threw it at the glass. The rubble disintegrated into a million pieces and barely scratched the window. She grabbed a loose metal pole, a spare one from the barricade, and clenched it in her fists. With all the remaining strength in her arms, she swung the pole over her shoulder and struck the window in the bottom corner. The shatter was loud enough to be heard by passers-by. A swarm of moths flew out, fluttering their hairy wings in Jasira's face, causing her to squeal and brush them away.

The window sill was filled with tiny sharp glass pieces which snapped, crackled and popped under Jasira's boots as she climbed through. The house was completely empty – no cardboard or plastic lining for beds, no single armchair in the corner of the room, no paint on the walls, no walls at all, but there were a few posts holding up the roof. All that remained in the house was the carpet in Vaika's old room and the timber framework that divided the two rooms. Without the walls, her old home looked much larger, and Jasira thought she should have considered removing them a long time ago, but she enjoyed the privacy back then.

There were no hidden places for Jasira to even look for the box. The only place Vaika could have put it was on the bedside table. *What a waste of time*, Jasira thought. *Who was I kidding?*

She climbed back out the window and made her way through the main walkway, stopping at the half-built central tower.

"What's the purpose of this tower?" Jasira asked one of the workers.

"No idea, love. The Cerberi won't tell us," she replied.

"My guess is it's the new infirmary for the Procedure," a voice said from beside her.

Jasira turned and immediately recognised the woman who spoke. Her red curls hadn't changed, or the brightness of her light freckles, although she was a lot less chubby than in her earlier years.

"Knock me dead," she said in surprise. Her eyes lit up as if she'd seen a ghost. "Jasira?"

Jasira immediately turned the other way and hurried down the main walkway.

"Jasira," Alixem called and chased after her.

Alixem reached out and touched Jasira on the shoulder, causing her to trip over her own feet. Jasira quickly covered her head as bystanders gathered around them. She wasn't sure how many people recognised her, but if Alixem did, then others could have too. She quickly got back up and left, leaving Alixem on the spot.

The last time Jasira was in The Border, she was fleeing the city. There wasn't even a chance to visit Vaika's grave at the time. Most people didn't bury their loved ones. They were usually cremated, and their ashes were kept in ceramic jars. If they weren't cremated, then the Cerberi disposed of them. Vaika always wanted to be buried, at least that's what Prianaj always told her.

The graveyard was in a secluded part of the city on the eastern outskirts. It was tucked away behind older buildings that were run down and falling apart. Nobody lived in them; nobody ever would. Vaika's mother was buried there, in a lonely plot with weeds growing out of the ground. Her name, 'Sirena Revken', was engraved on the headstone, which was mostly eroded. The date of death was illegible, so Jasira scratched out '2259' on it. Most of the graves were unmarked. Their wooden headstones had mostly been eaten away by termites or eroded by the elements. Vaika's grave was directly next to her mother's. It was a patchy plot, with barely any grass.

Jasira knelt down, placed her hand on the ground and rubbed her fingers into the dirt.

"I wish you could see him. My boy's all grown up," she said, sniffling. "Vi, I never thanked you for taking care of me for twenty-one years. You did everything to make sure I was safe and healthy. And I remember the time I didn't want to eat that terrible stew. I mean, it was terrible. So you went out in the middle of the night and stole some chicken and potatoes from the Cerberi for us to eat," Jasira continued, laughing and wiping the snot that dripped from her nose. "I never said goodbye, and I never thanked you properly. So thank you, sister, may we meet again."

"That was heart-warming," Alixem said from a distance as she applauded.

Jasira jumped up and wiped off her tears.

"How long have you been standing there?"

"I just got here. I figured you'd come to see your sister. I haven't seen you in seven years, Jasira. Your hair has changed, but it doesn't look like you've aged a day. Where have you been?" Alixem asked, walking towards Jasira.

"I moved to Monday."

"And didn't come back?" Alixem asked, walking closer to the grave.

"I had nothing left for me here. I'm only here to say goodbye to my sister," she said, pausing to tighten her jacket. "I didn't get to the first time."

"I remember your sister invited me over one day. We were going to make burgers. You were too young to remember," Alixem began.

"I think I remember that."

"You sister was so excited," she continued, shielding her eyes from the sun. "She made all the burgers and asked me to help carry them on a tray. I tripped over the carpet and dropped them all. I can't even tell you how bad I felt, but your sister and I laughed it off. She always saw the positives in every moment."

"She really did. That carpet was the death of me too. I can't even count how many times I tripped over it."

"I know," Alixem said, laughing. "There was always that loose bit near the corner that never stayed down."

Jasira's eyes widened as a light bulb clicked in her head. She hugged Alixem before running off.

"Thank you, thank you," she said, elated.

"You're welcome? Wait, Jasira," Alixem called out as she followed her.

Jasira's backpack was still under the steel barricade where she left it. She climbed through the window at the back of the house again and, with no time to waste, inspected the ground. She dug her nails into the carpet and scratched the floor. Dust flew into her face until her fingers stopped at a seam between two sections of the carpet. Jasira ran her fingers on the line in the carpet and noticed that it was seamlessly merged together. She tightened her grip under the smaller section and pulled it towards herself. The melted plastic under the carpet had already been broken. The harder she pulled, the louder the carpet ripped and cracked. Eventually, she pulled it loose and uncovered a door to a compartment underneath.

Jasira lifted the metal hook, and the air that was trapped underneath blew out into her face. However, it wasn't as stale as she expected. She held her nose and shut her eyes tightly, turning to avoid getting any more dust blown into her face. Once the air had cleared, she poked her head into the compartment under the trapdoor hoping to find something, but it was empty.

"What are you doing?" Alixem asked, jumping through the window.

"My sister had something. I remember she kept it when she was still alive. I thought it would still be here. I guess not."

"What was it?"

"Some sort of box. I never saw it in detail."

"Was it about... this big?" Alixem asked, moving her hands around to describe its size, which was no bigger than a small book.

"Roughly. Have you seen it?" Jasira responded, closing the hole back up.

"Cerberi raided most of the homes at this end of the city. I saw them holding a metal box a few days ago. They kept it pretty secret, they didn't want anyone to know."

"I have to get it."

"Wait, Jasira," Alixem said, pausing. "Can we talk about what happened between us? We were so close back then."

Jasira paused and recalled memories she had long forgotten. They were memories she had repressed because she couldn't deal with the emotions they brought.

"Alixem, I don't know why we drifted off, and I'm sorry we did. I really liked having you as a friend. I guess these things just happen to some people. You were always closer to Vaika than myself."

"I didn't want it to happen to us. I'm partly to blame, my mother didn't want me hanging around this side of the city."

"I understand her way of thinking. I'm sorry you had to deal with that. Let's sweep it under the rug now and put it behind us," Jasira said as she reached in to hug Alixem. "So, do you know where the Cerberi kept the box?"

"I overheard them saying that they would bring it to the masquerade. They must think it's important. It just looks like a stupid piece of metal to me."

"It's not stupid. I need it. This masquerade, when will it be?"

"Tonight," Alixem answered.

"Okay, we're going!"

"The Cerberi aren't just going to let us in. People like us don't get into those types of parties. They're exclusive, and your name will have to be on the list."

"Aren't you kind of... high up? You've been to these parties before. Surely, you'll be allowed in."

"Shit happens, Jasira. My family's high status is no longer recognised. We won't be allowed in. I know some friends on the wealthier side, but I doubt they'll help us."

Jasira placed her hand on her chest and felt Elaine's necklace press into her skin. She thought about the necklace for a second and grinned.

"They'll let us in. Trust me."

Idira unwrapped the heavily blood-stained cloths on the table and ordered one of her subordinates to call for Samaari. She inspected the contents carefully, moving pieces of dirt around with her baton.

Samaari pulled out a single strand of silver hair from her head as she walked through the corridors, examining it briefly before letting it float to the ground.

"This better be good, Idira," she said.

"Your majesty," Idira replied, pointing to Elaine's corpse. "Just as you asked."

"Bravo. Where did you find her?"

"Monday. Witnesses saw three people burying her body."

"Get me their names," Samaari demanded as she inspected Elaine's neck. "She's not wearing her necklace."

"We have one of their names, your majesty," Idira began, wrapping Elaine's corpse with the cloths. "Your aunt, Prianaj."

"Find the rest and bring them here," she said, walking back out.

"Your majesty," Idira called out, slightly raising her voice.

"Is there something else?" Samaari asked, pausing.

"The witnesses reported a child."

"Something wrong with a little girl?" Samaari asked.

"That's the problem. Witnesses claimed the child is not..."

"Not what?"

"The child is not female."

ELEVEN

MASQUERADE

ALIXEM CLAWED HER WAY THROUGH one of the old fallen skyscrapers with Jasira following close behind. The building fell on an angle and they crawled across giant glass windows to the other side. The glass beneath them were the only things between them and the city below.

"Where are you taking me?" Jasira asked, trembling on all fours and trying – but failing – not to look down. The ground below seemed a thousand kilometres away. The longer she stared, the farther down it stretched.

"Well, I assume you didn't bring any fancy clothing with you. We can't show up in ripped pants and faded shirts," Alixem replied as she looked back.

Jasira regained her composure and continued following Alixem.

They crawled underneath furniture that was crammed in the hallways until they reached an open space in a tower that penetrated the fallen skyscraper.

"Where are we?" Jasira asked, wiping dust off a counter with her finger.

"I found it a couple years back. You know this city is super boring, so I made my days a bit more adventurous," Alixem said, laughing as she disappeared into one of the rooms.

Jasira walked around the hall, pausing to look at everything she saw. She noticed a dusty dining room with plates, cutlery and chairs that hadn't been used in years. The chairs looked more plush than anything she had ever seen.

"Jasira come look at these," Alixem called out.

Laid out across a bed were dresses made from the most expensive fabrics in The Border. The only place Jasira had seen such extravagant clothing was the market hall in Monday. Back then, she'd never dreamt of being able to afford such expensive items.

"This is what I'm going to wear," Alixem said as she picked up a short black dress and a pair of high heels.

Jasira walked to the other side of the bed and ran her hands over two of the dresses. She quickly withdrew her hand from the first one, which was prickly. The second was soft and light, but she didn't like its yellow colour.

"This one feels nice, but it—" She noticed a bright red dress in a wardrobe across the room. "This one," she said, taking it off a rack and examining it. The dress was open at the back and had two strings that tied it together.

"That one will complement your skin tone," Alixem said. "You know what you need?"

"What, what is it?" Jasira asked eagerly.

"Shoes to go with it," she answered, holding out a pair of red high heels with straps. "And you need earrings."

"No. No piercings. My sister had one, and it always got infected."

"Oh, come on, it'll be fun. I know someone who can do it real quick. It won't hurt at all," Alixem insisted as she tried on her black dress.

"Fine. But don't make me regret this."

"What about mine? Do you think it's too tight?" she asked.

"No, that looks perfect on you," Jasira answered.

Alixem's friend lived in a small brick and mortar house on the wealthier side of The Border. It was tucked away between some nondescript buildings in an obscure part of the city.

Alixem knocked on the door twice, paused and then knocked again.

"My friend never learnt to speak, so you might find it a bit hard to understand her," Alixem warned. "And, don't stare."

"I'm sure it'll be fine."

The door opened slightly, and a face appeared from the shadow.

"Fenix, are you doing any piercings today?" Alixem asked.

The door opened wide, and they walked in. Jasira saw the woman's face and stepped back. It was covered entirely with piercings. The woman looked Jasira over from top to bottom.

"Piercing for who?" she asked.

"Me," Jasira said, quivering.

"You are Alixem friend?" she asked, moving to the living room.

"Yes, she's with me," Alixem said. "I think maybe an ear crawler will suit her."

Jasira sat on a velvet couch while Fenix opened a case that contained different types of ear crawlers. Fenix lifted the top layer to reveal a second set of ear crawlers that were much larger and shinier.

"Which one you like?"

Jasira took her time to examine each earring carefully. Her hand hovered over a set of snake and dragon earrings, but she changed her mind and quickly picked up a silver ear crawler designed in the shape of an animal fang.

"Do you have this one in gold?" she asked.

Fenix lifted the layer of ear crawlers to reveal another set of earrings. She picked up the same fang design in gold and placed it in Jasira's palm.

"This is the one," Jasira said.

Fenix pulled out a tiny wooden device that had a metal pin sticking out of it.

"This hurt little bit, but not a lot," Fenix said.

Jasira closed her eyes and clenched her teeth.

"We count to three," Fenix said as she rested the device around Jasira's ear. "One." She squeezed the device and poked a hole into Jasira's earlobe.

"Ouch! What happened to two and three?"

"Two, three."

Fenix wiped away the blood that dripped from Jasira's earlobe and quickly fixed the fang crawler to her ear. Then she pulled out an old hand mirror. Jasira turned her head to the side and moved her hair out of the way.

"This is beautiful, Fenix. Thank you," she said as she stared at her reflection.

"That one does look good on you, good choice, Jasira," Alixem said.

"What do I owe?" Jasira asked, still staring at her ear in the mirror.

"No worry. You Alixem friend. I give you free," Fenix answered.

"Where did you learn to how to pierce?" Jasira asked.

"I pierce Cerberi body, maybe, twenty Cerberi."

"How did you even get that close to them? They're so protective of themselves."

Alixem dropped a pile of jewellery she was examining and picked some rings off the ground. Jasira noticed her shove them in her pocket and return the rest to their container.

"I was Cerberi, long time ago," Fenix began as she closed the case of ear crawlers.

"Why did you leave? Cerberi are treated very well in Diektra," Jasira asked.

"You ask too many questions," Fenix continued. "But my leader, I no trust. Something very fishy about her."

Jasira twitched slightly and leaned back.

"Do you remember your leader's name," Jasira asked, hoping it wasn't Prianaj.

"Queen sister, Prianaj," Fenix said.

Jasira got up from the couch and thanked Fenix for the piercing.

"No worry, Jasira. Come back for more piercing next time," Fenix said, smiling.

"Wait, Fenix," Alixem said as she ducked under a chain that stretched from one wall to the other. "We need masks. It is a masquerade after all. Do you have anything?"

Fenix walked over to the bookcase opposite the couch and pulled it towards her, revealing a restricted section. She grabbed two face masks from the compartment and handed them to Alixem.

Jasira immediately grabbed the white porcelain mask and stared at it with wide eyes. It had red lips and was decorated with black rhinestones.

"All right, I guess that one's yours," Alixem said, holding onto her domino mask.

As night fell, the wealthy citizens made their way to the masquerade. At Jasira's old home, she and Alixem dressed up in their exquisite gowns and put on their masks. Alixem let her hair down past her shoulders while Jasira tied hers up. They put on perfume from Alixem's collection. This was the first time Jasira had worn perfume. Nobody ever traded it at the market.

Jasira strapped a dagger to her ankle for protection and made sure Elaine's necklace was visible on her neck.

"That's pretty, where did you get that?" Alixem asked.

Jasira hesitated for a second and then repeated what Elaine had told her. "Family heirloom."

Alixem decorated her hands with the rings she stole from Fenix's place and handed Jasira two of them.

"I saw you take those from Fenix," Jasira said.

"She stole them from me, I just went to get them back," Alixem said, laughing.

The masquerade, like all other events, took place in a private location underneath the city, 'The Clandestine', they called it. The Clandestine had three tunnel-like entrances, which were all hidden in clever spots around The Border.

Alixem and Jasira made their way to the closest entrance. Alixem knew its location because of the previous events she had attended.

There was a long queue of people dressed in fancy attire. Some of their dresses were quite long, and the women tripped over them. The queue moved fast, and it didn't take long for Alixem and Jasira to reach the front.

At the front, 'The Clandestine' was engraved into a plaque on the floor. The Cerberi greeted them and asked them for their names.

"Alixem Alice," Alixem said.

"I'm sorry, but your name is not here. Alixem, I believe your family is no longer welcome here," the Cerberi said.

Jasira coughed, tapping her chest. Her rings clinked against the necklace and caught the Cerberi's attention.

"Apologies, your highness," the Cerberi said, opening the door for both of them. "Enjoy your night."

"What was that all about?" Alixem asked.

"I have no idea," Jasira lied.

They followed the metal staircase down into the tunnel. It was dark, but they could hear noises that grew louder as they walked down the stairs. As they reached the main entrance to the masquerade, Jasira realised the noise they heard was music.

The hall was enormous, bigger than any room Jasira had ever seen. It was packed with so many people, who were all wearing masks. It was only the Cerberi who didn't have any masks on, but their mouths were covered as usual.

There were drinks on tall tables around the hall, and a live band was playing up-tempo music to lift the mood. They performed on a small stage that was not too far off the ground.

Jasira scanned the room looking for the box. People were crammed into a large circle near the centre of the hall. She followed the crowd and found, to her delight, the box staring at her. It sat in the centre of a black circular table and was guarded by a single Cerberi.

Jasira squeezed through the crowd and attempted to touch the box. The guard stretched her arm out and blocked Jasira.

"You can look at the box, but no guest is permitted to touch it," the guard said.

Jasira pointed at her necklace.

"Apologies, your highness. I didn't realise."

Jasira grabbed the box and walked towards the exit. Another Cerberi whispered something into the guard's ear before yelling out to Jasira.

"Hang on there," the Cerberi called out.

The band started playing faster, and the women danced on the dancefloor, twisting and spinning around the Cerberi. Jasira lost her balance and sense of direction. One guest clutched her arm and spun her around; another shoved her to the side. Soon, she found herself lost in an ocean of fast-moving bodies. The band continued to increase the tempo, and Jasira noticed the Cerberi making their way towards her.

Jasira ducked and dodged the Cerberi's punch. The women immediately dispersed to clear the dancefloor. The music stopped. Jasira jumped over a few tables and ran towards the exit. The Cerberi came at her from all directions with their batons held high.

"That necklace doesn't belong to you. We're going to need it. The box as well," one of the Cerberi said.

"I can't let you have it," Jasira responded. "And I don't think you're allowed to touch it. It's protected under Elaine's authority."

Jasira turned and made her way towards the exit. She sensed a Cerberi behind her and immediately threw her dagger in that direction, landing it in the Cerberi's forearm.

Alixem followed Jasira outside and reached for her shoulder.

"I have to go Alixem. Thank you for helping me with all this."

"Stay at least, this is a party," Alixem insisted.

"I've got no time. I only came get this box. That's it," Jasira said. "You go and try to enjoy the party; I've already ruined it."

Alixem hugged Jasira and wished her a safe journey back to Monday.

On her way to the front gates of the city, Jasira examined the etchings on the box. The lid was stuck, so she tried to pry it open with her fingers, but it was of no use. She looked around the box for a button or a latch but found only an engraving on the lid. It was the same shape as Elaine's necklace. Jasira noticed a tiny hole in the centre of the engraving that was the same size as the gem on the necklace.

Jasira hid in a tucked-away spot between two buildings and removed the mask from her sweaty face. She then took off the necklace and held it in front of her. The red gem glowed in the dark. She brought the gem close to the engraving on the box's lid, and it was pulled into place magnetically. Jasira heard a low click sound and the lid popped open.

The box was filled with letters. Jasira took out the first letter and immediately noticed the names of the sender and recipient.

Prianaj,

The citadel is exactly how you described it, dark and smelly. But it's definitely something I can get used to. Your sister, Marya, has had two children since you left: Samaari and Elaine. You probably already met Samaari, though. They only just found out that Marya raised Elaine in secret. She is being questioned by Queen Estella every day. I don't know what will happen to her, but I hope it's not anything terrible.

Elaine is the only friend I've made. Samaari doesn't talk to us much; they're training her for royal duties. Elaine and I try to sneak into the lessons, but we keep getting in trouble. It makes sense, though. I'm only a servant in this place. I like how Elaine makes me feel young again, like a child.

The food is absolutely insane, so much better than anything we ate in The Border. You probably already knew that ha-ha. The Cerberi still don't know where you've run off to, and they won't hear anything from me. I overheard your mother, Queen Estella, asking questions about you. I think she still cares. But other than that, there's not much going on over here.

How did it go with the garden you started?
Is anything selling well at the market?

Save something fresh and tasty for me.
I want to try it when I come back in two weeks.

Give my love to Vaika.

Best wishes,
Sirena Revken

Jasira covered her mouth with her hand.

"I didn't know Vaika's mother worked at the Citadel," she whispered as she packed all the letters back into the box and locked it. She wrapped the necklace back around her neck and put her mask back on before making her way to the front gates.

Jasira quickly stopped back at her old home to grab her backpack and change into more comfortable clothing. She arranged Alixem's clothes, which were in a messy pile on the floor, and left her rings on top of them. She then sneaked out and left the city through the front gates, unnoticed by anyone.

CHAPTER

TWELVE

LOST, FOUND, AND STOLEN

As Jasira returned to Monday, she noticed hundreds of Cerberi in the streets all over the city inspecting every passer-by. She dodged a few that headed in her direction, hiding in the shadows and backstreets.

From a distance, she recognised a woman who used to sell random stuff at the market – cutlery, sewing tools, threads and occasionally old books. Although, people rarely bought anything from her as her merchandise was usually antiquated. The woman was in her early seventies and was short and thin. She was a humble woman who always asked Jasira how her day was and tried to make small talk. However, Jasira was never able to remember the woman's name. A Cerberi threw the woman to the ground and ripped her jacket off. They did this to many random people in the streets.

What's going on? Jasira thought.

Jasira rushed to the apartment and knocked furiously on the door. Nobody answered. She knocked again and again, and there was still no response.

"Prianaj!" she called out, banging harder on the door.

Jasira knocked on the door so hard the hinges started to come loose. As she was about to take her last strike at the door, it swung open.

"Jasira. Come in quick," Prianaj said, tugging on Jasira's shirt.

Adam hid in the corner of his bedroom, and his door was closed.

"What's happening?" Jasira asked.

"I thought you were one of the Cerberi. They've been going to everyone's home, door to door."

"What do they want?"

"I don't know. But I have a hunch they're after Adam," Prianaj began.

"I think they want the necklace. They came after me back in The Border. But you might be right too. People would have seen Adam last week when he ran out."

"But I could be wrong," Prianaj continued. "I've never seen them this aggressive before. Not even when I was their leader."

"No, it makes sense they're after Adam as well."

"We can't leave until there's fewer of them out there."

"As long as Adam stays in his room, and we stay here, I'm sure we'll be fine."

Prianaj sat on the armchair and let out a quiet grunt as she let her body relax.

"The trip was great, thanks for asking. It was faster than I thought it'd be," Jasira said sarcastically, chuckling to mask the fear in her voice.

"Apologies, Jasira. We have other things to be worried about. But I'm glad you made it back safely. Especially with all those Cerberi out there."

Jasira pulled out the metal box from her backpack.

"Where did you get that?" Prianaj asked, getting up from the chair.

"It's my sister's box. The one I went back for. You've no idea the effort it was to get this."

"And for good reason. That box didn't belong to her. It belonged her mother."

"One of the letters inside was written to you."

"You opened it? How did you open it? It can only be opened with—"

"The necklace," Jasira interjected, revealing it underneath her shirt.

"Those letters were not for your eyes. All those letters were written to me. There's a reason Vaika couldn't open it. I should have destroyed that box years ago."

"You didn't tell me Sirena worked in the citadel. Why was she there?"

"You read the letters, it explains why. She was a servant for a few months," Prianaj said, hesitating.

"Don't lie to me. I've known you for a long time now. I can tell when you're lying."

Prianaj turned her head away from Jasira, pressing her fingers against her mouth.

"The truth?" she asked.

"The truth," Jasira repeated, sitting on the floor and holding Prianaj's hand.

"When I was banished from the citadel, I was pregnant. The Cerberi took my child by orders of Queen Marya. I never saw my baby again. When I came to The Border, I met Sirena, and over the years, we grew close. An opportunity came up for Sirena to work in the citadel as a servant for a few months. Together, we hatched a plan for her to try to find out what happened to my baby. She wrote to me all the time, and I wrote back."

"There's no way this is months worth of letters," Jasira said.

"Towards the end of her stay, Cerberi caught her writing to me," Prianaj continued. "I was banished so writing to me was not permitted, especially if they were still looking for me. And for good reason, I had the third necklace, and Marya wanted it. So they destroyed the remaining letters."

"I thought you told Elaine you gave your necklace to your baby."

"When Sirena came back, she told me my baby wasn't in the citadel. So, I gave the necklace to her so that if I was ever captured, it would protect her. I moved to the mountains and rarely came into the city, except to sell my produce. But before Sirena died, she told me that she found and gave the necklace to my baby. And that is all I know."

"That doesn't make sense, though. How did Sirena find your child elsewhere and not at the citadel?"

"I'm not sure. She never told me. And I didn't want her to tell me. I didn't want to be tempted to find my child."

"You didn't want to find your own child?" Jasira asked.

"If I had been caught with my child, who knows what Marya would have done to us."

"But the necklace. You would have been protected by it because your child would've had it."

"I feared what the Cerberi would've done. What Marya would've done. And I still fear what the Cerberi would do if they found me now. Especially if they found Adam," Prianaj said, quivering.

"Don't say that. They're not going to find him. I won't let them," Jasira said as she opened Adam's door.

He was crouched in the corner, his arms wrapped around his legs and his head buried in them. The noise from outside could be heard much louder in his room than in the other rooms. There was still chaos outside, and the women's muffled screams scared him. Every time he heard someone scream, his body shook and convulsed with fear.

"Adam, honey. Are you okay?"

As soon as Jasira spoke, there was a knock at the door. Cerberi called out for someone to open the door. Jasira's first instinct was to protect her boy. Without any second thoughts, she shut Adams's door, leaving only the two of them in one room, and Prianaj in the main room. She sat down beside Adam and covered his mouth.

"Adam, we're fine. We're okay."

"Open up!" Someone called from outside.

"Jasira," Prianaj whispered from the main room.

"Don't open the door. If they come in, don't tell them we're here," Jasira replied.

The Cerberi kicked down the door and forced themselves in. Prianaj's voice faded out as Cerberi dragged her away. Jasira felt Adam's breath get warmer on her hands, and she felt her heart thumping loudly in her chest. The voices in the main room were muted, so she crawled as close to the door as possible without making any noise.

"Prianaj. Long time no see," one of the Cerberi said.

"What do you want?" she cried.

"Elaine visited you, did she not?"

"I don't know what you're talking about."

"Don't make this difficult. Where's the necklace, Prianaj?"

Jasira grasped the necklace in her hand. *They're not after Adam, thank God.* She thought.

"Like I said, I don't know what you're talking about," Prianaj repeated.

"Queen Samaari doesn't want to make this hard. What about your necklace?"

Prianaj didn't move or say a word.

"All right then. Bring her," the Cerberi said to one of her fellow guards.

Prianaj struggled and kicked back, but she was too feeble.

"How dare you touch me. I'm a princess."

"Marya took that title away from you."

Jasira looked back at Adam, who started to feel the beginnings of a sneeze.

"No, Adam, don't," she whispered, but it was too late.

The Cerberi heard Adam's sneeze and kicked down the door to his room. They stared in disbelief at the boy, their jaws nearly touched the ground and their eyes almost popped from their heads. Adam screamed, covering his ears with his hands. He'd never seen Cerberi before, let alone aggressive Cerberi. Through the legs of one of them, Jasira noticed Prianaj unconscious on the floor with her head facing away from her.

"I don't believe it," one of the Cerberi said. "You were right."

"I know. I told you she was here, Bexi," Alixem replied, entering the room from behind the Cerberi. "Hello, Jasira."

"What are you doing with them?"

"Jasira?" Bexi repeated. "So you're the one who caused a bit of trouble in The Border. You fooled the Cerberi pretty well. Get the boy, and bring the mother. Be careful, we don't know if he bites." Bexi commanded.

The other Cerberi reached for the boy, but the closer she got, the louder Adam screamed.

"Alixem, why?" Jasira asked as Bexi restrained her.

"Get your boy to stop screaming, and I can explain," she said.

"Adam. Adam, calm down. Breathe. Stop screaming."

His voice dropped, and he brought down his hands.

"I followed you here. You were acting really suspicious when you first arrived in The Border. At first, I was concerned, but then something hit me like a brick in the face. They offered me my high status back if I helped them. They just wanted the necklace, Jasira."

"I'll hit you like a brick in the face. You did this just to get your status back? I thought you wanted our friendship back. That's what we talked about," Jasira said.

"Jasira, just take the necklace off, and they won't hurt you."

"But they'll take Adam. I can't let that happen."

One of the Cerberi dragged them out of the room, but Adam flung his legs around, trying to break free from her grasp. Prianaj regained consciousness but struggled to get up. Jasira saw the disappointment in her eyes; Prianaj was disappointment in herself. Jasira pushed Bexi from behind and shoved Alixem out of the way. She felt a blow to the back of her head and dropped to the ground. She was still conscious and saw Bexi crouch beside her.

"We've been looking for you for a very long time. It is quite a feat to stay on the run for seven years," she said.

Jasira looked in Alixem's eyes and emotionless face, nothing seemed to phase her. Adam and Prianaj screamed as the other Cerberi restrained them, and Jasira fell unconscious.

She struggled desperately to breathe as icy-cold water flooded over her. Her throat burned as if she had swallowed hot coals. Her body swirled around as she cried for help, but only bubbles of air came out of her mouth. She opened her eyes, but everything was a blur. Deep, red water swallowed her whole. She tried harder to escape the dream this time, but ever since they moved to Monday, it felt more real.

CHAPTER

THIRTEEN

THE BEAST

JASIRA'S CHEEK PRESSED AGAINST the cold stone floor. Her hands were covered with bruises, and her eyelids were heavy. The room she was in was almost too dark to know if her eyes were even open; it just bright enough for her to see the vague forms of the objects in front of her. But she couldn't focus because her eyes ached as if they'd been plucked out and reinserted in their sockets.

Prianaj groaned in the space next to Jasira, pounding the ground with her fist. Jasira tried to rush to comfort her, but the metal bars between them blocked her. The bars glowed red with heat, and she quickly moved away from them before she burnt herself.

"Prianaj, where are we?" Jasira asked as she backed away from the bars.

Prianaj grunted as the soft glow of the bars travelled through the entire dungeon. Suddenly, the bars appeared behind her, above her and in every direction she turned. She was trapped in a cage; the entire room was filled with endless rows of cages each glowing soft crimson. Prianaj was in the one beside her, and a stranger in the next one down.

"It's the dungeon. We're in the citadel," Prianaj said, pushing herself off the ground.

Jasira immediately thought about Adam. She called out his name but didn't get a reply. In the cage opposite hers, there was a young woman who appeared to be in her late teens. Her face was smeared with dirt, and her hair was bushy.

"Have you seen a little boy, about this high?" Jasira asked the woman, raising her hand to indicate Adam's height.

"A boy? Never in my life. Not in my nineteen years on Earth," she replied.

"Prianaj, do you know where they might have taken Adam?"

"He could be anywhere; the citadel is huge. I don't think they would have locked him in here with us. Samaari would have questions for him. And questions for us too."

Tears streamed down Jasira's cheeks.

"I've gotten him killed," she sobbed.

"No, Jasira. Don't think like that. They won't touch him before they talk to us."

"You're right," she said, wiping her face. "Besides, he's protected under Elaine's authority."

"Not exactly."

"What?"

"He's only protected if he has the necklace. Since he doesn't, Samaari doesn't know of the protection vow."

"I have to get this to him then," Jasira said, pulling out the necklace.

"Put that away. If the Cerberi see you with that, it's over for us all."

"What do they want with it anyway?"

"There's something very important that can only be unlocked with all three necklaces. When the Cerberi attacked us back home, they demanded I hand over Elaine's necklace. They wanted mine as well. Obviously, I had neither, so they took us. Samaari must be collecting them to open it."

"What do they open?"

"It's like that metal box, the one with all the letters. The lock on the lid can only be opened with the necklace. There's a vault embedded in the wall behind the throne. With all three necklaces, it can be opened. But nobody has ever opened it.

When I was still in the citadel, we had all three necklaces in our possession, but we basically forgot about the vault. There was no reason to open it."

"What's inside it?" Jasira asked.

"I'm not entirely sure. They say that whatever is in the vault can be used to destroy but it cannot be destroyed."

"What does that even mean?"

"We never questioned it," Prianaj continued. "Diektra was doing just fine before all this."

"I still have to get the necklace to Adam. But where would they have put him?"

A metal door opened, scraping against the stone walls. The noise echoed throughout the dungeon, rattling the cages. The other prisoners hushed their voices and made little movement. Cerberi entered in pairs. They pressed a button on a machine similar to the generator Jasira destroyed years ago. The electricity passing through one of the cages in front was turned off. Its bars stopped glowing, and one of the Cerberi unlocked it. From a distance, Jasira recognised the Cerberi, Bexi, the one who captured them.

"Come now. Time to go," Bexi said to the prisoner.

"Where are we going?" the prisoner screamed in fear.

"Guess we'll find out, won't we?" Bexi said as she dragged the prisoner out of the cage by her legs.

The other Cerberi grabbed the prisoner's arms, and together, they lifted her out of the dungeon with the door slamming behind them. The other pair of Cerberi walked up and down the gaps between the cages, hitting their batons against the bars. Jasira didn't know if they were trying to scare the prisoners or just mocking them. Nevertheless, she knew she didn't want to be there anymore. Besides, just like in Sirena's letter, the citadel was dark and had a foul odour – in the dungeons at least.

The Cerberi laughed in the faces of the prisoners, mimicking their cries. They were wicked and had no self-control. One of the Cerberi moved to Jasira's cage and scratched a knife on the bars.

"Where's my son?" Jasira shouted, leaping to her feet.

"This one talks," the Cerberi said, laughing. "Aye Frayda, this here is the mother of that creature."

Frayda struck her baton against Jasira's cage.

"You shouldn't have spoken to them," Prianaj said.

"You're just as insignificant as the others in here," Frayda said to Prianaj. "Hang on. Do my eyes dare deceive me? I haven't seen you in a long, long, long time," she said, laughing. "Prianaj, if I remember correctly. You were banished, you no longer have authority here."

"What happened to all you Cerberi? I never trained you this way," Prianaj asked with disappointment in her voice.

"Times change, don't they? Charley, bring the mother. The beast might be hungry, and the queen will want to see Prianaj," Frayda said.

Jasira moved to the back of the cage.

"Beast?" she asked, quivering and drenched in sweat.

"It gets hungry," Charley whispered.

The two Cerberi restrained Jasira and took her through tight corridors illuminated by fire-lit torches. They stopped in front of a door, and Jasira heard a low growl coming from behind it. Frayda unlocked it with a fancy-looking key and threw Jasira inside. Everything happened too quickly for Jasira to fight back or even scream.

The room she was in was warmer, probably from the heat of the sunlight coming through the gaps in the wall. Smeared across the floor and walls was a sticky red solution. Jasira felt it between her fingers and instantly knew it was blood. She could see a river down below through one of the gaps in the wall. The only other parts of the citadel she could see were the roof and a courtyard much further down below. She couldn't see much else, but she noticed the room she was in was very high above the ground.

There was a part of the room that was completely dark. Something growled, slowly. It poked its head into the sun, and its menacing black eyes reflected the light. Jasira stayed close to the wall, her back flat against it. The beast's sharp claws scraped against the floor as it emerged from the shadows. Its mouth opened to reveal diamond white fangs. A breeze entered the room through the gaps in the wall and flowed across the beast's golden mane. Jasira's heart pounded so hard that she thought it was about to jump out of her chest. She pushed her legs against the floor, but there was nowhere to go. She squealed. The beast glanced at her. Jasira closed her eyes and thought about every wrong decision she'd taken that had led to her current circumstance. The beast inched closer to her, but it didn't touch her – not yet.

Jasira slowly opened her eyes and saw the beast bowing in front of her. She exhaled in relief but was still unsure if she was alive. She saw the beast in all its glory; it was a lion. It breathed heavily into her face. She recognised the animal from when she saw one in The Border years ago.

Jasira stretched her arm, and the lion stepped forward and brought its head into her palm. She rubbed her fingers through its soft, silky hair. Then the lion moved its head and pressed its nose on Jasira's jacket. She felt something hard in her jacket that she wasn't sure was there before. She removed a small book from her pocket and realised it was the Bible Vaika had gifted her. She turned to a random page and read the first verse she saw.

'My God hath sent his angel, and hath shut the
lions' mouths, that they have not hurt me:
forasmuch as before him innocency was
found in me: and also before thee,
O king, have I done no hurt.' – Daniel 6:22

Just as God sent his angels to protect Daniel in the lion's den, Jasira was also saved from the beast.

She bashed on the door for the Cerberi to open up, but they didn't respond. They were waiting for the lion to attack Jasira, but they didn't know she had been spared. The lion walked in circles around Jasira and crouched at the door, letting out a thunderous roar. The noise reverberated through the door, and the Cerberi opened up. Jasira stood tall with her shoulders back and head high. The lion stood beside her, ready to pounce.

"The queen needs to see this," Frayda said, looking at Charley in confusion. "Follow us."

Jasira followed them up a tall flight of stairs with the lion by her side. Prianaj was being taken in the opposite direction by another set of Cerberi.

Before Jasira reached the top of the stairs, she noticed Adam kneeling down between two guards. She immediately ran to him and threw her arms around him. The guards tried to stop her from touching her son, but Samaari waved her hand at them to allow it. The lion stayed close to Jasira, moving in circles around her and Adam.

"You're okay. Oh, Adam. I'm so sorry this happened," she said.

"I don't believe it," Samaari said as she rose from the throne.

Jasira looked around quickly and her eyes scanned the room. She bowed and moved to the centre, just opposite Samaari. She peered to the side to see if Prianaj was telling the truth about the vault behind the throne, but there was a large curtain that stretched over it. For the first time, Jasira got the chance to see the queen. She had piercing silver eyes and deep black hair. She was tall and thin; it was almost as if her skeleton could be seen through her pale skin.

"Amazing, isn't it?" Samaari continued.

"I'm sorry, I don't understand."

"The lion, my beast, was supposed to devour you. Yet, here you are, untouched and unharmed," she said, tossing the lion a chunk of meat. "Jasira, isn't it?"

"Yes, your highness. I'm so sorry for all th—"

"No need to apologise. You brought me your greatest weapon. I should be thanking you."

Jasira paused and raised her eyebrows. Her eyes widened, but when she opened her mouth to speak, no words came out.

"You have no idea what you are," Samaari said as she called the lion to her side. "Don't worry, the beast won't attack me.

You see, they used to call lions the kings of the jungle. Dangerous and deadly predators, but still in control of their kingdom. They bring no harm to those with royal blood. Who are you, Jasira?"

"I'm just a regular person. I swear. I'm from The Border."

"Who are you really?"

"I never knew my mother."

"Don't lie to me. You were found in Monday," Samaari said as she slapped the back of her hand across Jasira's face. "Elaine found you for a reason. I would ask her but, you know, she's dead."

"We moved to Monday after I gave birth," Jasira said, touching her reddened cheek. "With all due respect, your highness, Elaine wasn't looking for me. She had been looking for my son for seven years."

"Why would she be looking for the boy," Samaari asked as she regained composure.

"I'm not sure. The only thing Elaine said before she died was that the boy is proof," Jasira said as she flinched every time Samaari made a movement.

"Proof of what?"

"I don't know that either," Jasira said, looking towards the ground.

"And how long have you known Prianaj?"

"She lived in The Border and has helped me out ever since I had my son."

Samaari sat back on her throne and began to play with a dagger, spinning it around her fingers.

"You know, Jasira, I can't let the boy live."

"No, no, please. Elaine, he's under the protection of Elaine," Jasira said, pulling out the necklace.

"Give that to me," Samaari demanded.

"I won't let you take this," Jasira said, placing the necklace around Adam's neck. "Elaine protected him, and you can't touch him now that he has the necklace. And you can't take the necklace off him, he'll have to take it off himself."

"Adam don't take off the necklace, okay. Promise me you won't," she implored Adam.

"Promise," he replied, nodding.

"Elaine trusted me, and she believed that Adam was destined for something great. So I will honour her wishes, and you have to as well."

"Take that thing off him this instant," Samaari said, her veins bulging in her neck. "Who do you think you are to do this?"

"My name Jasira Revken. A nobody from The Border."

Samaari took in a deep breath and exhaled slowly. Her tone immediately changed.

"Take the lion back to its den," she said to the Cerberi. She then stood up and opened the double doors on the left wall. "Take Jasira to my quarters."

CHAPTER

FOURTEEN

THE CITADEL

S AMAARI'S QUARTERS WERE GINORMOUS, almost bigger
than the throne room, except for the fact that her quarters
had two levels. It also had the same tall ceilings as the rest of the
rooms in the citadel.

Jasira explored the lower level first, starting with looking
through the wardrobe. Samaari owned hundreds of garments.
Jasira pulled out a stunning silky blue dress and laid it on the
bed. She wasn't worried about being caught; Samaari needed
her if she wanted the necklace. Samaari had different styles of
the same garment; some were winter clothing, but most were
loose summer clothing. One of Samaari's dresses alone looked

more expensive than everything Jasira had ever owned – more expensive than anything she had ever seen in both cities. If she could go back in time to the masquerade, she'd take a few of Samaari's dresses with her to wear.

Jasira stood in front of a cracked mirror on the wall and held up the blue dress against her body, admiring herself in it. The dress looked too long for her. Samaari was a very tall woman, and all her garments were tailored to her height. Jasira imagined living like a royal, although she would never recognise the woman staring back at her in the mirror.

Jasira took off her shoes to relax before she moved to the upper level of Samaari's quarters. One entire wall was cut out, letting in an abundance of fresh air. Of course, there was a railing to stop anybody from accidentally falling over the side. At the top of the staircase was a pristine white baby grand piano. Jasira pressed a few keys, and the sound was so crisp she got goosebumps. On the floor was a large plush rug that felt comfortable and soft under Jasira's feet. She looked over the balcony, but there wasn't much to see. Samaari's quarters faced the Pacific Ocean, which looked endless. The breeze was cool and refreshing. She enjoyed the sound of the waves; somehow, she felt calm and free.

Jasira walked back to the lower level to find a servant waiting for her wearing a cream-yellow dress and a bonnet over her head. She kept her head low, even when talking.

"Don't mind me," she said while cleaning around the room.

Jasira helped put back some of the clothes she had taken out of the wardrobe.

"You don't need to help me. I'm fine on my own," the servant insisted.

"No, it's all right. I took them out. I can help."

"You remind me of someone who used to live here."

"Really?" Jasira immediately thought of Vaika's mother. "Was her name Sirena?"

"No. But Sirena was kind as well, a servant like me for a few months. I mean the queen's daughter, before Marya banished her. She wasn't permitted to keep her child, especially because... no, I shouldn't be telling you this."

"Are you talking about Prianaj?" Jasira asked.

"Yes, did you know her?"

"Yes, she was brought here with me," Jasira said, following the servant upstairs.

The servant dusted the top of the shelves and readjusted the furniture. Jasira sat on the piano stool and rested her hands on the keys, accidentally making a sound.

"Do you play the piano?" the servant asked.

"No, I've never learnt. But I would like to hear someone play one day," she replied.

The servant sat on the other half of the stool and gently placed her fingers on the keys. She pressed down and allowed her hands to dance on their own across the piano keys. Jasira closed her eyes and allowed the music to move her body.

"I shouldn't be doing this," the servant said, stopping the music.

"That was really beautiful. How did you learn?"

"It came naturally to me. I could understand sheet music from old books."

"You'll have to teach me one day, if I'm still here," Jasira said, chuckling. "Back to Prianaj, why didn't they let her keep her child?"

"I can't say."

"I won't tell," Jasira said.

"When Prianaj was banished, she had just given birth. The queen at the time kept the baby here in the citadel while Prianaj was kicked out," the servant began as she folded clothes and placed them back on the shelves. "I never saw the baby, but there were rumours floating around that the baby was... different."

They walked back to the lower level.

"Different how?"

"I don't know. We were silenced, we couldn't speak about it. But one night, the baby was stolen."

"By Prianaj?"

"They want to believe so. But I don't think it was her. I think somebody who worked inside the citadel took the baby.

The Cerberi would have caught Prianaj if she stepped foot inside the citadel."

Jasira sat on the bed and processed the information. She remembered Prianaj telling her that Sirena didn't find her baby in the citadel.

"Why would someone take Prianaj's baby?" she asked.

"That's all I can say, I'm sorry."

"No, please," Jasira pleaded.

"It has something to do with a curse. It's the only thing that can explain why two preceding queens broke the law to bear two children, and Samaari's stillbirth. That's why she's gone so crazy. Please don't tell the queen I said that," the servant answered.

"Do you think that's why Prianaj was banished? It doesn't really make sense."

"I mean, maybe. Prianaj was banished after she had her baby, there had to be another reason as well. But she still stayed in Diektra. She ran to The Border. I would have thought she would run to another kingdom."

"Diektra is the only remaining kingdom in the world. All the others died out after the Culling."

"That's what they want us to believe," the servant said as she made her way to the exit.

A Cerberi walked in as the servant opened the door.

"Samaari will see you now," she said to Jasira.

The Cerberi escorted Jasira back to the throne room, where Samaari waited. Jasira followed Samaari through the double doors to the left of the throne. Water dripped onto a wide stone bridge that stretched to the mountains beyond the citadel. The water came down from a waterfall that was spectacular and much higher than any cliff Jasira had ever seen. It splashed into the river below, which ran into the Pacific Ocean.

"This view is definitely something."

"I come here to think most of the time. To clear my mind. Being queen is not easy," Samaari said. "Let me ask you, now that we are alone, did Elaine say anything else about the boy?"

Something in Jasira told her not to reveal anything.

"No, nothing else," she lied.

"You wouldn't lie to me, Jasira. You know I'm the queen."

"I know. But I'm not."

"Prianaj didn't say that. She told me everything my sister said about me, the citadel and the kingdom."

"It's not much of a kingdom really. The two cities of Diektra are separated, fear runs rampant through each of them and nobody has seen you or any member of the royal family, well, ever. Do you even know what's going on in your kingdom?" Jasira asked.

"I know more than enough," Samaari said as she moved closer to Jasira and spoke right into her face. "Now take that necklace off your son!" she said, raising her voice.

"That necklace isn't coming off him."

"There won't be another chance like this."

"I don't know why you so desperately need the necklace... for yourself or the vault—"

"How do you know about the vault?" Samaari interjected as her brows drew together. "Ah, Prianaj."

"This was never about Adam or stopping the male species from returning. All you wanted was the necklace." Jasira realised.

"Just take it off him. I want it now!" Samaari said as beads of sweat rolled down her temples. Flecks of spit landed in Jasira's face as she spoke.

"That necklace is never coming off Adam," Jasira said, confidently.

"Then you leave me no choice, Jasira. Give my regards to Elaine," she said as she shoved a dagger into Jasira's abdomen and tossed her body over the bridge. Samaari screamed as she looked up at the sky as it seemed to fall away from. In her anguish, she pulled on her hair, almost ripping it out from the roots.

Jasira struggled desperately to breathe as icy-cold water flooded over her. Her throat burned as if she had swallowed hot coals. Her body swirled around as she cried for help, but only bubbles of air came out of her mouth. She splashed her hands around and opened her eyes, but everything was a blur.

Her ankle scraped on gravel and rocks, and she bled profusely. Clouds of crimson oozed into the rushing waters. Deep, red water swallowed her whole.

PART III

CHAPTER

FIFTEEN

AEDYN

I DIRA'S FOOTSTEPS WERE HEAVY on the floor as she paced towards Samaari. Her fellow Cerberi greeted her as she walked past them. She, however, kept her head high and ignored them.

"The boy is in the dungeon," she said to Samaari.

Samaari didn't turn to acknowledge her. Instead, she stood facing the window on the right side of the throne. She stared at Monday in the distance. The terrain between the city and the citadel was a wasteland, barren and useless.

"Until the boy takes off that necklace, he doesn't eat," she said.

"Why don't we just feed him to the beast, and you take the necklace once he's dead?"

"That won't work. Levon won't attack the boy." Samaari had named the lion. "It didn't attack the boy's mother, they share the same blood, so it won't attack the boy. If only you had half the brain my sister had."

"Apologies, my queen. But I don't think the boy will take the necklace off. He has covered his ears, he won't listen to us."

"Like I said, no food until he takes it off. Don't feed Prianaj either. He needs to see who's in charge. If only I hadn't recited the oath when I ascended the throne, I would've ripped that necklace off him so fast his mother would've had come back from the dead to stop me."

"Back from the dead?"

"She wouldn't comply. I threw her off the bridge."

"I don't think you should have done that," Idira said.

"Are you questioning my authority?"

"Forgive me your majesty." Idira said as she stepped back and looked down.

"Don't tell the boy. If he knows, then he may never give up the necklace. And you know how desperately I want it."

"Can't I take the necklace off him?" Idira asked.

"And curse this citadel even more? Curse us all?"

"A thousand apologies," Idira said as she walked out, moving through the corridors to the kitchen.

Thankfully nobody was there, so she took small pieces of bread and hid them in her pocket. She then walked through the corridor to a flight of spiral stairs that led to the back entrance of the dungeon.

"Out of my way," she said to the two Cerberi at the door. She sounded calm, but her heart was racing.

"There are no prisoners scheduled for removal today," one of them said.

"I am your superior. Don't question me. But if you must know, I am doing a routine check-up," she replied.

"As you wish," the other Cerberi said, opening the door.

Idira looked through each of the cages for Prianaj. She found Adam with his hands over his ears, looking away from her. Idira took one of the pieces of bread and tossed it into his cage. He brought his hands down, picked up the bread and nibbled at it. His face looked grim in the dark. Idira smiled warmly, and her heart softened as she continued to search for Prianaj's cage.

"Prianaj," she whispered loudly.

A cough came from one of the cages in the centre.

"Prianaj," she repeated.

Somebody coughed again from the same direction. Idira followed the noise to the cage and saw Prianaj sitting on the ground. She tossed her the second piece of bread.

"Samaari wants the necklace," she said.

"Why are you telling me this?" Prianaj asked.

"I don't want any of you hurt."

Prianaj looked away from Idira.

"Look, I know my loyalties have shifted over the years, but after realising Samaari's plan, I can't stay on her side. I just can't."

"Empty words, Idira."

"When you taught me, decades ago, you told me something that stuck with me. Ever since you left, the Cerberi have not had a better leader. You told me and I quote, 'In this world, we fight for what we believe is right', and I have stuck by that. You should have been the one to ascend the throne. So you're going to regain your rightful place. I'll make sure of it."

"I don't want the throne."

"But we both know Samaari doesn't deserve it."

"So what's your plan? To sneak up on Samaari and dethrone her? You were my strongest soldier but not always the smartest."

"I have to get you out of here first, then we take the boy to safety."

"How do I know you aren't lying?"

"One thing you know about me, Prianaj, is that I'm not a liar. In all my years as a Cerberi, now a leader, I have never lied, and you know it. That's got to mean something," Idira said, accidentally touching the bars and singeing her fingers.

"All right. But you need to make sure the boy's mother is safe with him. I don't want to see her caught up in this mess."

"Prianaj," Idira said softly. "You know how crazy Samaari is. She doesn't think before she acts."

"Idira. What is it?" Prianaj asked, worried.

"Samaari... killed her. She threw her off the bridge. I didn't know, I swear. She only just told me."

Prianaj trembled. Her lips quivered, and her eyes blurred. Tears ran down her cheeks as she sobbed, covering her face with her shaky hands. Her throat tightened as she struggled to take in deep breaths. It felt as though her heart had been ripped out of her chest and her soul were torn in half. She shrieked in pain, sobbing as she tried to process the news.

"Just go. Leave!" she screamed.

"I'm so sorry, Prianaj," Idira said, walking back out of the dungeon while wiping a single tear from her face.

Jasira coughed up water as she lay facing the sky. Someone applied pressure to her chest and breathed into her mouth. Her cold fingers dug into the sand around her. Her vision was blurry, and she could only see the silhouette of the person leaning over her body. The last thing she remembered was being lifted off the wet ground before she lost consciousness.

Jasira replayed in her mind different moments of her life. The deeper she got lost in her thoughts, the more she couldn't differentiate between what was real and what was not. Strange noises from outside interrupted her, growing louder and louder until they finally snapped her out of her reverie.

Jasira opened her eyes and immediately shot up from the bed. Her wet hair soaked the pillow and most of the mattress. She was in a small aluminium shed with a skylight. It glowed a soft white, revealing somebody sitting in the corner of the room reading a book.

"Where... where am I?" she asked, forcing the words out.

The person closed the book and placed it on a bench. Jasira noticed the person's short hair and sharp bone structure. For seven years, Jasira had thought Adam was the only male in the world. As the person moved closer to her, she could see him clearly. He had silver eyes, tanned skin and bushy black hair. He was a bit older than Jasira, in his late thirties.

"You washed up on the beach. If I hadn't pulled you out of the water, you would be dead," he said.

Jasira squinted her eyes and tried to recall how she even ended up in the water. Then she remembered what Samaari did to her, feeling for any wounds on her abdomen, but there was no pain there. She reached for her jacket and noticed the dagger was still stuck in it. The fabric was torn where the blade pierced the jacket, exposing the white padding. The dagger hadn't pierced Jasira's body but was wedged into her Bible.

"You're lucky to be alive."

"Who are you? What's this place?" Jasira asked.

"Aedyn. This is my home."

"Have you always lived on your own?"

"For the most part, yes," he answered.

"I need to get back to the citadel. How far is it?" Jasira asked.

"You washed up pretty bad. I stitched your ankle, and your body is bruised. The citadel is quite a journey, up north."

"My son is there. I fear for his life."

"Another male," Aedyn said in excitement.

"How much do you know, about the rest of the world?"

"A woman took care of me in my younger years, she told me stories of bad people. She said that the world had gotten rid of men and that I would never be safe enough to return to the mainland."

"That's true enough. I thought my son, Adam, was the first boy."

"How old is he?"

"Seven."

"I remember when I was that young. If feels like a lifetime ago."

"I still can't believe you've been living here your entire life. With no people or—" Jasira looked at the table opposite her and saw a dead fish, "without food," she said slowly.

"Oh, don't mind that. That one's old. I only eat the fresh ones."

"Right."

Jasira swung her legs off the bed and attempted to walk. She cried as she felt intense pain throughout her body. Her ankle was bandaged in a clean white cloth, and her hands were grazed.

"Don't get up, you're hurt pretty bad," Aedyn said as he put his arms around her and helped her back into bed.

As he leaned over her to fix the sheets, Jasira caught a glimpse of his necklace. She reached into the top of his shirt and pulled it out. This one was silver with the same mark as Elaine's. There was a sapphire in the centre tied to a black twine.

"How do you have this?" she asked.

"The woman who took care of me gave it to me," he said.

Adam has Elaine's necklace, and Samaari would have hers. The only other necklace is Prianaj's. The thought ran through Jasira's mind.

"Do you know your mother?"

"I never met my real mother. The woman that took care of me always said that I could never meet her."

"What about the woman that took care of you, what was her name?"

"I can't remember. It was too long ago. Her name is on the tip of my tongue. Sophia? Sarah?" he pondered.

"Sirena?" Jasira asked.

"That's it. Did you know her?"

"My sister's mother. But that necklace of yours belonged to someone I know," she said, letting him go. "It couldn't be... no, Prianaj would have told me her child was... I remember her telling me her baby was a girl... she said her baby was stolen... no, she didn't say dead... she would have known... she should have... did she?" she whispered aloud.

"You need to rest. By the way, I'm sorry, I didn't get your name."

"Too many people know my name," she said.

"Well, it's nice to meet you. It's nice to finally talk to someone again," he said as he left the room.

Jasira mumbled in her sleep, tossing and turning at the sound of crashing waves outside. The aluminium walls were so thin that every sound from outside could be heard clearly. She woke up in the middle of the night in a cold sweat, screaming. Her entire body was soaked. Aedyn rushed into the room with a lit candle. Jasira sighed, hoping it was all a nightmare.

"I heard you screaming," he said. "Are you all right?"

"My son is all alone, probably fearing for his life in that citadel. What if they take him to the lion? He'll be terrified," she cried.

"There's a lion?"

"More than a lion. When I saw your necklace, it didn't hit me at first, but the longer I thought about it, the more I realised I must have known your mother," Jasira began.

"No, the woman who took care of me gave me this necklace, not my mother."

"Your mother owned it. She gave it to Sirena to give to you," she continued. "There are three. My son has the first. You've got the second one, and the queen has the third. Only people of royal blood have them."

"Are you—"

"No, I'm not royalty. The queen's sister, Elaine, gave me hers before she died. It's a long story. But your mother was royalty."

Aedyn took off his necklace and admired it in the candlelight.

"I'm a prince?" he asked.

"Don't get too excited."

"What's her name? My mother."

"Prianaj," she said.

"Prianaj," he repeated.

"If you want to meet her, she's in the citadel. But you've got to help me get there. And she has a lot of explaining to do."

"I've never been to the citadel, but it seems like quite the journey."

"We have to leave as early as first light."

"I don't know how dangerous it will be. You're still weak, you need time to recover," he insisted.

"I'm strong enough. I've been on dangerous expeditions before. But you're right. Let me recover tonight."

Aedyn crouched under the bed, pulled out a black case and placed it on the edge of the mattress. He popped open the two latches on each end and pulled out a titanium sword. It was so shiny that the candlelight reflected off it like a dazzling kaleidoscope. The edge of the sword was sharp, and the hilt was wrapped with a white rope. Hidden inside the lid of the case were two short daggers and an array of throwing knives.

"Where did you get those?" Jasira asked, admiring them with bright eyes.

"Sirena brought them so that if anybody found me, I would be able to protect myself," he answered.

"How well can you defend yourself?" he asked.

"You've no idea what I can do."

"If you really think you're strong enough, we leave tomorrow. I've always wanted to use these, for protection obviously," he said, laughing.

"Tomorrow, at first light."

"At first light," he repeated.

"Perfect. But I get to have the little knives," she said, chuckling.

Aedyn pulled out a coastal map from the case. He placed it on Jasira's lap and pointed to the wavy markings.

"Now the hardest part won't be the dehydration. It'll be getting through this, the rapids. I've built a small boat, but I'm not sure if it will last. We can row across the beach to the mainland on the other side. It might shorten the trip by a few hours. You came through the rapids, and it's no doubt a miracle you survived, but it'll be hard going against the current."

"Is the boat big enough for both of us?"

"We'll have to squeeze into it, but we'll fit."

"What if we take the boat across to the mainland, and then walk the rest of the way once we reach the other side? It might slow us down, but it would be easier than going against the current."

"That sounds like a better option. Glad you washed up on my shore."

"I'm glad I didn't die."

Aedyn laughed.

"Sirena must have struggled to get to you all those years ago. She definitely hid you out of reach from other people."

"She cared a lot for me. She always wanted the best," he said in a soft voice.

"Well, better get some rest. We have a big day ahead of us, and we want my ankle to heal if we are to leave at first light."

As Aedyn packed up the sword and daggers, Jasira reached for the map but grabbed his hand instead.

"Aedyn," she said. "Thank you for saving me."

He quickly pulled his hand away. He paused at the door and looked back at Jasira. He was about to speak but changed his mind and walked away.

CHAPTER

SIXTEEN

THE FERRYMAN

BLACK SAND COVERED THE BEACH ENTIRELY, and foamy waves crashed on the shore. The tide was high, and cold winds blew in from the ocean. Large, dark clouds formed in the sky.

Aedyn pulled the boat from a separate shed and let it float in the water. It was as small as he'd described it, and Jasira was unsure the both of them would fit inside. The wood of the boat looked worn and in desperate need of maintenance. But there was no time for that.

"Are you sure this will hold out there?" she shouted over the loud winds as she pointed to the rough waters.

"I wasn't expecting the water to be this bad. It looks like a storm is brewing. But trust me, the boat will hold."

"How do you know it won't break?"

"Because I built it. We have to leave now, before it gets worse," he shouted.

Jasira helped Aedyn push the boat off the sandbank, and she jumped in. Her ankle had healed quickly overnight, but whenever she applied too much pressure on it, she felt the pain again. Aedyn moved the sword case out of the way to make space for Jasira and handed her a hefty leather belt.

"Tie this around your waist," he said as he pulled out the daggers.

He held them close to the belt, and they magnetically stuck to it.

"Wow. That was cool," Jasira said.

"It might take a few tries to get used to taking them off and putting them back on the belt, but once you do it the first time, you'll get the hang of it," he said, attaching the rest of the daggers.

He hung a leather scabbard over his shoulder and attached the sword magnetically to it like the daggers. The case was now empty, so he placed it in the water and pushed it to shore. Jasira moved the two oars out of the way to sit more comfortably and handed them to Aedyn.

"Let's go save them," he said, handing one back to Jasira.

Together, they rowed the boat closer to the mainland.

"Yesterday, why did you refer to Sirena as your sister's mother?" he asked, breaking the silence.

"Sirena took me in when I was young, since before I can even remember," Jasira began.

"Oh, I'm so sorry. What happened to your mother?"

"I don't know anything about my real mother. I never met her," she continued. "I grew up with Vaika, Sirena's daughter. She was basically my sister, and we treated each other like family. I never stop thinking about what my life would have been like if my biological mother raised me, what it would have been like if I never met Vaika."

"Where's your sister now?"

Jasira shook her head.

"She... died. A long time ago, before Adam was born."

"Oh, I'm—"

"It's okay," she interjected. "Everywhere I've been, there's death. Everyone who I come into contact with dies. Elaine, Sirena, Vaika."

Aedyn gulped and widened his eyes.

"Don't worry, Aedyn. We have your weapons to protect us. We'll just be in and out of the citadel. We grab Adam and Prianaj, and we leave. My whole life I never thought I'd be doing something like this. I'm not sure Sirena would like the person I have become."

"Sirena really cared for people. She took care of me, and she took you in as one of her own. She wouldn't think that of you at all. I don't think that of you."

"You've known me for one day, and you think you know me? Aedyn, you don't know me at all. I've done terrible things. I've killed people to get where I am, to keep my son safe," she said, staring out to the ocean.

"Are you listening to yourself? To keep your son safe," he repeated. "You protected him. And now, look, we're on our way to save him."

"Well, maybe there's a darker part of me. A part of me I never knew existed, just like my life before Sirena took me in."

"Just choose which life you want to live," he said.

"It's not that easy. You've lived alone. You haven't had the same experiences as me."

"For good reason. If the world is really that terrible, then Sirena was wise to keep me away from it. Yes, I've been alone since I was ten, but I've had choices to make too. When Sirena didn't return, I could have lived my life on the beach or tried to make it to the mainland and probably drowned. And if I made it, then they might have killed me. I had to make that choice as a little boy. You've known me for one day as well, and you don't know me either."

The blanket of fog cleared up a bit, and they could see the water had settled down. There was a crack in the clouds that allowed the sun to break through, and the winds abated.

"The storm will return. This usually happens," Aedyn said, taking Jasira's oar and rowing a bit faster.

"Slow down for a second, I like this calm."

The part of the water they had rowed into was dark blue. The sunlight didn't travel far enough to reach the bottom.

"This water reminds me of a story Sirena once told me. It's similar to the situation we're heading into and also involves a boat," he said as the dark clouds covered the sun again.

"I haven't heard a story in a long time. Vaika used to love making up stories and telling them to me before bed. But there was always one story she never wanted to tell me. Apparently, I was too young and would be too scared," she mimicked Vaika's high-pitch voice.

"What was it called?" he asked.

"She called it 'The Ferryman'."

"The exact story I was thinking of."

"Can you tell it to me?" she asked in excitement.

"I don't know if you can handle it. Do you like ghost stories?" he asked.

Jasira stared at him, and he instantly knew that she wanted him to tell the story.

"Fine, but don't complain when you get nightmares."

"I don't get scared easily," she lied.

"There was once a woman who had a daughter who was eight years old," he began as the swoosh of water accompanied his voice. "She loved her daughter with all her heart and would let nothing come between them. One day, as the woman was out, her home ignited into flames. Her daughter was inside. That very day, the woman lost her child, and her whole world fell apart. She began to lose her mind and fell into a dark place. She found a witch and begged her to bring her daughter back to life, but the witch would not. The witch forbade it. She did, however, direct the woman to somebody who could."

"The ferryman?" Jasira asked.

"The ferryman," he repeated. "The woman followed a path to a lake in the middle of a dark forest where nobody ever dared to enter. She rowed the boat into the very centre of the lake and waited. Time went by, and she lost hope. But right before she started to row back to shore, black smoke emerged into the shape of a hooded figure before her eyes. It was the ferryman. The woman pleaded for him to bring her daughter back from the dead. The ferryman agreed but asked for payment. He asked the woman to give up something she could never live without. The only thing the woman could offer was her daughter, the very thing she wanted the ferryman to bring back. So the woman told him to take whatever he wanted from her. But the ferryman was devious. If he brought somebody back from the dead, the other souls would become restless and see it as an unfair deed. His one and only job was to ferry souls from this world to the next. So the Ferryman took the woman's eyes, gouging them out and swallowing them. He vanished, and the woman's daughter appeared in the boat. The woman was so excited when she heard her daughters voice again but could not see her. She reached out to give her daughter a hug but couldn't reach her. She leaned further forward with her arms out and fell into the lake. She fought to save herself but could not see the boat to pull herself up. And so she drowned, without her daughter and without her eyes."

A chilling wind blew across Jasira's skin and rocked the boat slightly.

"What happened to the woman's daughter?" she asked.

"When the ferryman took out her eyes, he deceived her and implanted a memory of her daughter in her mind. The woman didn't realise her daughter hadn't returned from death."

"Are you my ferryman? Are you going to take my eyes?" she asked.

"It fits the scene, doesn't it? We're on a boat, in the middle of, well, not a lake, but an ocean, and you want your child back."

Jasira ate the nuts and berries they collected earlier in the morning. Aedyn had learned to grow produce the same way

Prianaj, his mother, did. They had been rowing for a couple of hours, but the mainland was still out of sight.

Aedyn moved the oar up too hard and accidentally splashed water onto Jasira's face. He laughed, but Jasira didn't like it. She grabbed the other oar and splashed water onto Aedyn's face as payback. Aedyn continued laughing and splashed Jasira again. Jasira swung the oar into the water once more but hit against something hard.

As Jasira looked over the side of the boat, she saw a hand poking out of the water. She jumped back and screamed.

"What? What's it?" he asked.

Jasira covered her mouth and pointed. Aedyn used an oar to pull the body closer to the boat and flipped it over. Despite the pale skin, Jasira could recognise those red locks and freckles.

"Alixem."

"You know her?" he asked.

"Knew her," she answered. "She's the reason we're in this mess. She was an old friend of mine, but she turned me in and helped the Cerberi capture my son. I don't know her anymore."

Aedyn pushed Alixem's body away and let it float across the ocean.

"You think you know some people, then they stab you in the back like that. I never had any friends, the only person I knew was Sirena. Hey, do you know what happened to her? Why she never came back?" he asked, warmly.

"She died when I was really young. She's buried next to my sister."

"Do you know how she died?"

"My sister said it was an infection called the night patch. It's a horrible disease, not contagious, but some kind of genetic disorder. Vaika died from it as well."

"I always wondered why she never returned. Maybe when this is over, you can take me to where she's buried."

"Definitely. I think she'd like that."

A storm rolled in and rocked the boat with waves that increasingly grew stronger. Jasira held onto the sides of the boat as Aedyn fought against the winds, using all his energy to row. One of the oars snapped. It had it coming; the wood was terribly rotten.

"Look," he said, pointing behind her.

Jasira turned her head and squinted her eyes.

"The mainland," he said over the winds.

The waves grew stronger, splashing water into the boat and soaking them both.

"We have to swim the rest of the way."

"What! No way. I can't swim!" Jasira screamed.

"There's no choice."

One final wave crashed into the boat, breaking it apart into several pieces. They swirled around in the water, spinning as the current pushed them apart. Jasira struggled to stay afloat and noticed Aedyn clutching to a larger piece of the boat.

"Swim. Watch how I do it," Aedyn commanded.

Jasira watched him as he kicked his feet and pushed the water with his arms. She copied him, putting pieces of the wood under her body to keep her afloat. Her body was thrown around, but she fought for her life, to see her son again. Aedyn manoeuvred through the rough waters powerfully, as he already knew how to swim.

Jasira coughed up the water that tried to force its way down her throat. After struggling for a while, her legs gave in to fatigue. She thought it was over for her, but somehow, her legs kept moving. The pain in her injured ankle felt unbearable, but she knew she was close. The water was up to her neck, except when the waves rolled over her head. She pushed towards the mainland, swinging her arms around to help her body move faster. Aedyn made it to the shore much sooner and watched as Jasira crawled up on the shore.

"We did it!" she said, puffing and trying to catch her breath.

It rained heavily, and mud slid down the side of the mountain into the ocean.

"Take a rest, you're exhausted," Aedyn said, stabbing his sword into the sand.

Jasira checked her body for the daggers. She thought the waves might have knocked them off her, but they remained intact on the belt.

"Strong magnets these are," she said.

"Yeah, I know. I thought we would have lost the weapons out there."

"We need to get to higher ground," she insisted, handing Aedyn his sword.

They made their way up the mountain, slipping in mud and wet leaves. The rain was cold, but the wind had subsided. As they climbed to the other side, they saw the citadel hidden in the mountains. The waterfall close to the citadel was loud, and they could hear it over the drumming of the rain. It was hard to see the Cerberi at the edges of the citadel, as their black uniforms camouflaged with the blackstone walls.

"Let's wait until night," Aedyn said. "If we go now, we'll be caught."

Jasira moved some fallen branches and leaves out of the way so that the rainwater could pass freely. She sat down and squeezed the water out of her hair and jacket. They had soaked up more water than she thought possible. Aedyn sat next to her, his back against the tree and his legs outstretched in front of him. Jasira calmed herself with the song she always used to sing to her son.

'Close your eyes and lay down,
Dream a very sweet dream.
I'll be right here to stay,
You'll be safe when you're with me.'

Aedyn closed his eyes and continued the song.

"Close your eyes, my child,
In you, I'll find my strength.
I love your beautiful smile,
I'll be yours until the end."

"You know that song?" Jasira asked.

"Sirena sang that to me every night," Aedyn answered, wrapping his arms around Jasira's shoulder.

"My sister did the same for me," she said, resting her head on his chest.

They repeated the song over and over, swaying their bodies side to side in each other's arms. Aedyn's voice faded into a soft hum, and it wasn't long before Jasira's did too.

"I don't like the sky like this, I can't see the stars," Aedyn whispered.

"You like stargazing?"

"Sirena taught me. She showed me all the important stars. She kept a notebook of all the incredible things she saw, even before I was born. She left it to me, I have it back at my beach shed."

"I wish I knew how to stargaze," Jasira said.

"I can teach you one day," he smiled. "I went through her notebook, and she recorded so many amazing things. Like the day of my birth, the moon was dark crimson. Sirena always said it made me special."

Jasira spun her head to face Aedyn and remembered the prophecy.

The prophecy isn't mine, its Prianaj's. She realised.

She pondered on it for a while and realised that Aedyn was not to blame for Prianaj hiding the truth about her child.

"Jasira," she said.

"What?" Aedyn asked.

"My name. It's Jasira."

They were exhausted, but they only needed a few hours of sleep to replenish their bodies. They drifted off, ignoring the water that dripped on their heads.

SEVENTEEN

CONFESSION

THUNDER CRASHED AND LIGHTNING FLASHED in the sky, waking Jasira and Aedyn from their slumber. As they got up and prepared themselves, the rain grew heavier and heavier, and tree branches snapped under the force of the wind. The petrichor smell emanating from the ground was almost too strong to handle.

As they ran towards the citadel, the treacherous terrain and relentless downpour caused Jasira to slip and fall again and again, but she wouldn't be deterred as she knew Adam was within reach. Meanwhile, Aedyn was agile and manoeuvred easily through the forest, striding over fallen trees with his long legs and pushing through overgrown branches with his broad shoulders. Jasira was surprised by Aedyn's strength and power. She had never seen anyone, any woman, move that fast before

Jasira eventually caught up to Aedyn, who was waiting for her near one of the high walls of the citadel. Like most of the citadel, the wall was built of blackstone bricks. Aedyn cupped his hands and bent his knees slightly.

"Jump up and tell me how many guards there are," he whispered.

Jasira jumped up and tried to grab the top of the wall. She slipped, but Aedyn caught her. She tried again. This time, Aedyn boosted her by pushing her legs up. Once at the top, she lay flat, blending in with the wall.

There were no lights outside the citadel. The only part of the citadel that had electricity was the dungeons. The courtyard was illuminated by fire-lit torches that were arranged in rows on each wall. Jasira saw only one Cerberi walking around the courtyard.

"There's one," she said.

"Are you sure?" he asked.

Jasira looked again, surveying the courtyard more carefully.

"Positive," she replied.

"All right. Pull me up now," he said, jumping to reach her hand.

Aedyn used his legs to climb up the side of the wall as Jasira pulled him up. They lay flat on the top of the wall, facing each other.

"Okay, what's the plan?" he asked.

"Why do I need to come up with the plan?"

"Because you've been in the citadel before. I have no idea where anything is or where Adam and Prianaj are being kept."

"They're in the dungeons," she said.

"All right then. To the dungeons."

"It's not going to be that easy. The only male the Cerberi have ever seen is Adam. They'll try to kill you. I'll go get Adam and Prianaj, and I'll bring them here to you."

"You can't let anybody see you," Aedyn said.

"I'll need to be invisible."

"How are you going to be invisible?" he asked.

They both turned their heads to the Cerberi patrolling the courtyard.

"No. No way," he said.

"I have to. I can't let them stay any longer in there. What if they're torturing him to get the necklace?"

"That's understandable," he said as he nodded.

Jasira threw one of her daggers on the ground beneath them, and its clang drew the attention of the Cerberi.

"Who's there?" the Cerberi called out as she walked towards the wall, clutching her baton.

She noticed the dagger and picked it up. Jasira winked at Aedyn and rolled over to the side, landing on top of the Cerberi and knocking her out.

She struggled to fit into the Cerberi's uniform but eventually squeezed into it. She tossed her clothes up to Aedyn, who dropped them over the other side. Aedyn dropped a few more daggers to Jasira in case she needed them. She dragged the Cerberi's body to the corner of the courtyard where there was a door to a tiny storage room. She left the body inside and used a heavy flowerpot to block the door from opening.

"Good luck," Aedyn whispered aloud.

Jasira put on the Cerberi's mask and entered the citadel. She regained her bearings and moved through the hallways. She passed by a Cerberi who nodded to her, and she nodded back.

There were more different-sized doors on her right than her left. Soon, she realised she must have gone around the same hallway multiple times. A set of stairs appeared on her left, and as she moved closer, she heard Samaari talking to someone at the top but couldn't make out the words. Jasira recalled the path she had taken the last time to get to the throne room, and she backtracked. She moved through the hallways until she got to a door that looked very familiar to her.

"Beastie," she said as she gently placed her palm on the surface of the door.

She looked ahead and saw the spiral stairs that led to the dungeons. She quickly noticed the two Cerberi guarding the entrance.

"What are you doing down here, Orbina?" they asked.

Jasira looked down at the blue letter 'O' embroidered on her uniform. *That's what the letters are for,* she realised.

"Right. Um. Just checking on the prisoners. Samaari's orders. You know how she is," Jasira said, coughing.

"Is everything okay with you? You don't sound right," they asked.

Jasira cleared her throat and replied, "Just a cold."

"Okay then. Who's watching out front?"

"Uh," Jasira hesitated, trying to remember a Cerberi's name she'd heard before.

She looked at the letters on the uniforms of the two Cerberi in front of her, one with a green letter 'B' and the other with an orange letter 'U'.

"Orbina?" the Cerberi said.

"Um. Frayda took over," she said, coughing.

The Cerberi opened the doors and let Jasira into the dungeons. She immediately closed the door and flicked the switches on the giant machine, turning off the power to the cages. To her surprise, the cages were all empty. Not a single prisoner remained. The two Cerberi entered the dungeon with their batons out. They twisted the tip of the batons to reveal metal spikes.

"Orbina would have known the prisoners were all fed to the beast, except the old lady and the boy. Levon wouldn't eat those two. Who are you really?" they asked.

"I didn't know our batons did that," she said, running between the cages.

Jasira was felt too tired to be running, especially because of the ordeal she had gone through earlier in the day and her injured ankle. She huffed and held the sides of her waist. The mask didn't let in enough air. She couldn't breathe. The Cerberi cornered her between the gaps of two cages, one behind her and the other in front of her. They closed in on her and raised their spiky batons. Then Jasira heard the Cerberi behind her shout in pain as she fell to the ground.

"Idira?" the Cerberi in front of her said.

Jasira kicked the Cerberi's stomach and pushed her away. The Cerberi fell and knocked her head on the side of the cage.

"Who are you and why are you here?" Idira asked.

Jasira took off her mask. Idira's eyes widened, and she lowered her baton.

"I thought you were dead," she said.

"Dead women tell no tales. Why did you do that? Why did you help me?"

"The enemy of my enemy is my ally," Idira said.

Jasira put the mask back on, and they walked back through the hallways.

"I know where your son is. He's with Prianaj in a separate holding bay. Samaari ordered us not to feed them, but I couldn't watch an innocent child suffer. Samaari's crazy, but that's a bit too far."

"Thank you. That means a lot. Samaari will pay for all this."

"I can agree with that," Idira said, opening the door next to the lion's room.

This room was much like the lion's, cold and empty, except for the two people inside. Jasira dropped her mask and ran to her son. She hugged him tight and didn't want to let go. Adam held Jasira and cried. He'd never been away from his mother for such a long period.

"You're alive," Prianaj said. "But Idira you told me—"

"I thought so too. Until I found her in the dungeons, looking for both of you," Idira interjected.

"But how?"

"Your child saved me," Jasira said confidently. "On the beach."

"My child?"

"Your son. Why didn't you tell me? All those years and you kept that secret from me. I can understand why, but still, you could have trusted me. We agreed to tell each other the truth."

"Jasira, you more than anyone know how difficult it is for me to trust people," Prianaj said.

"You made me believe I was the special one, that my son was somehow important. The prophecy was never about my son. It was about you. You birthed a male child during a blood moon, and you are royalty. You knew this and still kept it from me. Why?" she asked, perplexed.

"I didn't want to believe it, I couldn't. That my son was the one the prophecy spoke about. You have every right to be angry, Jasira," Prianaj said reassuringly.

"I'm not angry, just disappointed. Was there even a blood moon when Adam was born?"

"No. I told everyone in The Border that a blood moon had passed. But I did that to protect you and Adam. I was protecting you from my past. I couldn't burden you with my own secrets. I cared for you both, loved you both, leaving my own son behind."

"Loved us? In this world and the next right. It's what you wrote to him, that letter. Yeah, I found it in the cave seven years ago. You didn't even look for him. Sirena had to take care of him for you. And when she died, he took care of himself. Don't sit there and tell me you loved us, because love isn't leaving your child scared and alone," Jasira said, taking in a deep breath and exhaling slowly. "He came with me to finally meet you.

He never knew you but yet has the courage to risk his life to have a second chance with you."

"He is here?" Prianaj asked, smiling.

"Yes. But if you want to meet him, we need to get out of here before anyone sees us."

Jasira opened the door to the lion's room and left it wide open. The lion's roar echoed through the entire citadel.

As they walked through the rest of the hallway, Jasira didn't speak to Prianaj. She just held Adam's hand close to her. They rounded a corner, and a Cerberi stopped them. Jasira held Adam tight with both hands.

"What are the prisoners doing out?" she asked Idira.

"Samaari called for them," Idira answered.

"Impossible. I was just with her."

"She called for them earlier. We got caught up," Idira hesitated.

"I'll come with you two. Orbina, I thought you were guarding out front," the Cerberi said to Jasira.

"I asked for her assistance," Idira stepped in.

Jasira looked at Idira, who shrugged her shoulders.

As usual, Samaari sat on her throne, high and mighty. For the first time in a long time, she wore the crown. It was a simple gold ring around her forehead.

"Why are the prisoners out?" Samaari asked.

"You called for them, your majesty," the Cerberi answered.

"I did no such thing. Idira, what's the meaning of this? You've gone crazy, just like my sister. Stupid girl."

"The enemy of my enemy is my ally," she mumbled.

"Come again?"

"The enemy of my enemy is my ally," she repeated, charging at Samaari with her spiky baton.

The lion immediately jumped in front of Idira and swiped its claws at her. Idira fell to the side and remained unconscious on the floor, bleeding from the exposed flesh in her arm.

"Good boy, Levon," Samaari said, fixing her dress. "That girl had no brains. Trying to attack me, who does she think she is?" Samaari said, laughing.

The lion walked to Jasira and bowed. Jasira placed her hand on the lion's head and bowed back.

"Orbina is not royalty, what's this?" Samaari said.

"You don't deserve that crown, Samaari!" Prianaj yelled.

"Quiet!"

"Let her speak," Samaari snarled.

Prianaj backed away.

"See, she has nothing useful to say. Take them back to the holding bay. Separate them this time," Samaari commanded.

As one of the Cerberi reached to grab Adam, Jasira sliced her face with a dagger, leaving a deep cut. She fell to the ground, holding her face as she screamed in pain. Jasira threw the dagger at Samaari, but she ducked out of the way and hid behind her throne.

"Orbina, what are you doing?" Samaari screamed.

"Run!" Jasira said.

Prianaj grabbed Adam's hand, and they ran to the front courtyard. Thankfully, there were no Cerberi outside, but a wailing siren sounded through the entire citadel. Adam covered his ears and cringed.

"You made it!" Aedyn shouted, jumping down from the top of the wall.

"Quick, we have to go now," Jasira said, almost slipping on the wet floor.

"Wait," Prianaj said, placing her hand on his cheek. "Is it really you? My son."

"Aedyn, this is Prianaj, your mother," Jasira said.

They hugged, finally reuniting after decades of separation. Cerberi quickly flooded the courtyard, holding up their spiky batons.

"Stop!" Samaari yelled. "You know I only want the necklace. Oh my, what do we have here? Another male. Prianaj, your son? Years back, nobody believed the rumours that your child was a boy, yet here we all are. Now all I want are the necklaces. Prianaj, I know you have yours, and I need the boy's one as well. That's all I want, and I'll let you go."

Aedyn took his necklace off and shoved it in his pocket. Unknown to him, the necklace's black twine hung out.

"You're lying!" Jasira shouted.

"Orbina, whose side are you on? We the necklace holders don't lie to each other," Samaari said, turning to face her Cerberi. "Get them!" she commanded.

"What is wrong with all you Cerberi? Never in my years of training you all would you follow this tyranny!" Prianaj yelled, furiously.

All the Cerberi slowly marched towards Jasira and Prianaj. Aedyn carried Adam away from the group to protect him. Levon jumped in front of Jasira and Prianaj and roared, saliva spattering out of his mouth. Jasira smiled in relief. The Cerberi all ran back into the citadel, fearing the lion's sharp fangs. Samaari walked towards Levon, waved her hand and then pointed behind her. Bound to Samaari's authority, Levon obeyed her. As the lion ran towards the edge of the citadel, it looked back at Jasira and nodded.

"No... no," Jasira cried. "No!" she screamed louder.

With a poker face, Samaari threw her dagger at the lion without even looking back at it. The dagger lodged itself in the lion's underbelly, and as it tripped over its feet, it fell over the cliff.

Adam took off his necklace and held it out.

"No, Adam, what are you doing?" Jasira yelled.

Adam released himself from Aedyn's arms and ran to Samaari.

"No more fighting," he said, handing it to her.

Samaari laughed. "Your boy has chosen his own fate," she said, pulling out another dagger.

Aedyn rushed to Adam's aid and grabbed him in his arms. Samaari took a swing and missed.

"Come on!" she screamed in frustration.

Aedyn ran with Adam towards the citadel's outer wall. Samaari squinted one eye, stretched out her arm and let the dagger loose. Aedyn's necklace fell out of his pocket and bounced onto the floor. As the dagger screeched and whistled through the air, it exploded into three separate daggers moving in the same direction. Adam fell out of Aedyn's arms and slid across the slippery pavement. Aedyn also fell to the ground with a loud thud. Samaari ran to Aedyn's necklace and ripped it out of the crevice it had lodged itself into. She held it in front of her face and laughed. She then pulled out the two daggers that had lodged in Aedyn's spine. The third dagger had bounced off his sword and landed on the ground.

Prianaj limped as fast as she could towards her son while cursing at Samaari, who ran back inside the citadel. Jasira followed Prianaj to Aedyn. Adam rubbed his grazed knee and then moved towards Aedyn's lifeless body. He shook Aedyn's shoulders, but he just rolled over onto his back. The blood from his wounds seeped onto the floor, mixing with the rainwater and staining the blackstone bricks.

"Aedyn, Aedyn, wake up," Jasira said, shaking his body as well.

The tears that fell from Jasira's eyes merged with the raindrops that were pouring down on her. She sobbed as she held his hand; Prianaj held the other and wailed.

"I just got you back, my son," Prianaj cried. "Please don't leave me now."

"Aedyn, please wake up," Jasira cried louder.

"Jasira," Prianaj said, shaking her head. "You know what to do."

Jasira nodded and kissed Aedyn's forehead. She rose with a throwing knife in one hand, Aedyn's sword in the other and a fire in her heart that burned with a desire for a vengeance the likes of which she had never felt before.

CHAPTER

EIGHTEEN

LOST, FOUND, AND TAKEN

A S SAMAARI REALISED SHE was being chased, she increased her pace through the corridors. Jasira ripped off the breastplate from her uniform and threw her mask to the ground. She followed Samaari down a long corridor, who clenched both necklaces in her hand. Finally, they were in the same hallway.

Applying all the lessons Prianaj had taught her, Jasira aimed a dagger at Samaari and then let it fly. It moved swiftly in a straight line and struck the back of Samaari's hand. She dropped both necklaces and screamed, pulling out the dagger and throwing it at Jasira. She missed, and Jasira ran towards her, but Samaari hurried up the staircase to the throne room. She didn't look back.

Jasira picked up both necklaces, wore them around her neck and followed Samaari.

"Samaari!" Jasira called out.

Rain splashed into the room through the open windows. The winds blew out the torches, and the curtains danced in the dark, casting long shadows of fear and terror into the citadel. The door to the stone bridge swung open and crashed onto the walls.

Jasira looked around and breathed in. She took small steps towards the door and hesitated. Her heart pounded in her chest, and she froze up. She held the sword with both hands and stretched it out in front of her. Samaari stood in the rain at the far end of the bridge, with two daggers in her good hand. She had tucked her injured hand under her armpit to stop it from bleeding out.

"Orbina! Idira tried to convince you that I'm crazy. But I'm not, I'm not crazy!" Samaari shouted.

She could only see Jasira's silhouette, but a sudden bolt of lightning illuminated the bridge. Samaari dropped her daggers, widened her eyes and shook her head. Her face immediately went pale, and her hands shivered.

"What's the matter, Samaari? Looks like you've seen a ghost," Jasira said.

"No. You're dead. I killed you."

"If you want me dead, you're going to have to try again."

"Fine. But let's make this a fair fight. No weapons," Samaari said.

Jasira threw her sword and daggers to the ground. Samaari did the same, taking out every hidden weapon in her dress. She had hidden them surprisingly well.

The rain fell harder, flowing over the bridge into the river below. Thunder crashed, and the sky flashed repeatedly. The bridge was slippery, but that was the least of Jasira's worries.

"If you want to be a hero Jasira, then you'll have to die like one!" Samaari yelled.

"I don't care about being a hero. Samaari I know about your miscarriage. I'm so sorry."

All the blood in Samaari's body rushed to her head, bright red like a volcano ready to explode.

"You don't know anything! My daughter would have been queen. Now there is no successor to the throne."

They charged towards each other, moving to the centre of the bridge. Jasira threw a punch at Samaari's face with all her strength. Samaari dodged it and struck Jasira in the stomach. She then kicked Jasira's injured ankle, painfully twisting it. Prianaj hadn't trained Jasira well to fight without weapons.

"Come on. Fight!" Samaari screamed, taking off her crown and throwing it to the ground. It clanked and rolled into the throne room.

Jasira fell to the ground, hunched over and cried in pain as her ankle bled.

She shook her head and said, "What's the point? You've taken so much already."

The sword scraped against the floor as Samaari dragged it off the ground.

"I'll make it quick. You won't feel a thing," Samaari said as she held it high above Jasira's head, staring at the back of her neck.

Jasira saw Samaari's reflection in the dagger on the floor in front of her.

"Give my regards to Elaine this time," Samaari said.

Jasira closed her eyes and recalled memories of her son, flooding her mind with thoughts and emotions. She pictured his bright, smiling face and wondered who was going to look after him after she was gone, especially if something terrible happened to Prianaj. After all, Prianaj was getting old. Aedyn was dead so he wouldn't be able to take care of Adam. Vaika and Sirena were dead too. Jasira clenched her fists and mustered all her strength. Something deep inside her heart fought to reignite her desire for vengeance.

Samaari took a swing at Jasira, but she ducked in the nick of time. The sword scraped a stone behind her. Only the finest hairs on the back of her neck were sliced.

"Argh," Samaari growled.

A flash of lightning revealed an empty space in front of Samaari. Jasira had disappeared.

"Come out and face me. Don't back away," she shouted.

A shadow ran past Samaari and swiped a dagger at her. It was too quick for Samaari to attack. The shadow dashed behind her. Samaari spun on the spot, breathing heavily. She was kicked and punched repeatedly, badly bruising her face. Jasira pushed Samaari to the edge of the bridge and kicked the sword out of her hand. Samaari managed to jump up onto the parapet of the bridge and crouched down. Jasira held the dagger to Samaari's neck.

"Go on. Kill me," Samaari said as she laughed with blood oozing from her mouth. "You don't have the desire."

"I'm not like you, Samaari. I'm not a monster," Jasira said.

Samaari pushed Jasira out of the way but slipped and lost her balance. Jasira reached for Samaari but could only manage to grab her bronze necklace. Samaari reached for Aedyn's necklace around Jasira's neck with her injured hand to pull herself up. She was shaky and couldn't hold on for long.

The black twine stretched and became very thin. Jasira's neck was in pain as it held Samaari's weight. Finally, Samaari's necklace snapped and she fell over the bridge, disappearing into the rushing waters. Jasira leaned over the bridge and shook her head, taking quick breaths. She glanced at Samaari's necklace in her hand and tied it around her neck balancing her head up as her body became exhausted. She sat down in the rain, stretched her legs and finally breathed a sigh of relief.

The sun broke through the dark clouds, and the rain stopped. Jasira had fallen asleep on the side of the bridge with a piece of fabric around her ankle to stop the blood. A Cerberi poked Jasira's shoulder with her baton, waking her. Jasira jolted up and grabbed her ankle, which was still in enormous pain.

"Where's the queen?"

"Dead," Jasira answered.

She saw Adam and Prianaj restrained by another set of Cerberi inside the throne room. Jasira limped towards them and picked up the slightly dented crown. The gold coating hadn't been scuffed, though. She took it to Prianaj and polished the crown with her clothing.

"There's no one else to ascend the throne. It's yours," she said.

"I'm not taking that. I've seen it destroy too many queens before Samaari."

"Who will take the queen's place?" one Cerberi asked.

Idira grunted as she pulled herself up from the floor. She held her head as it throbbed and squinted her eyes to regain her vision. She pressed a rag to her arm to stop the bleeding. Jasira took the crown to her.

"You're leader of the Cerberi. It's yours if you want it," she said, handing the crown to her.

"Thank you, but... this crown is powerful. I can imagine myself as queen. But saying no to it... is more powerful," Idira said, handing the crown back.

Jasira looked around at all the Cerberi inside the throne room. She looked at her son and smiled, then at Prianaj who nodded her head slowly. She recognised the servant who just walked in, the pianist. She took off her bonnet and nodded as well.

Jasira looked around the room again and limped towards the throne, spinning the crown in her hands.

"I will," she said as she placed it on her head and sat on the throne.

It was cold and heavy on her head, but she kept her back straight and bore it's weight. The throne was hard and cold against her back. She liked the instant feeling of power rushing through her body, power she had never felt before. However, she didn't know how long it would last.

"Leave us," she announced.

The Cerberi left the room, untying Prianaj and Adam. Adam ran to his mother.

"What now?" Prianaj asked.

"I will watch over the citadel until we can decide who will rule. Take Aedyn's body to The Border and bury him next to Sirena, he wanted to visit her grave when this was over. He'd like that. Don't announce Samaari's death yet," she answered.

Adam played with the curtain behind the throne and tugged at it too hard. It ruffled as it fell to the ground, revealing the vault. Prianaj didn't lie about that, at least. The vault was a large metal plate embedded into stone. Jasira pushed against it, trying to open it, but it was stuck. A smaller square plate popped forward, revealing a lock sequence similar to that of Vaika's metal box. There were three swirled grooves in the metal plate, one for each necklace. Jasira grasped the necklaces around her neck and widened her eyes. She discerned that when Samaari fell over the bridge, she took Aedyn's necklace with her. The vault could never be opened, even if they wanted it to. Jasira took off Elaine's necklace and placed it around Adam's neck.

"Adam, listen to me. Don't take this off ever again. You promised me last time, and you still took it off. People died for this, good people too. I won't let it happen again," she said, tying a knot in the chain so he couldn't fit his head through the loop.

"No more fighting?" he asked.

"No more fighting," Jasira repeated as she sat back down on the throne and crossed her injured leg over her other leg.

Idira sat on the floor and leaned on the side of the throne, with her hand on her head.

"I can't believe Samaari is dead," she said, laughing.

"Long live the queen," Jasira said, laughing with Idira.

As the waves crashed on the black sand beach, two older women walked along the shore with their toes digging into the sand. One of them had straight light blue hair and the other frizzled brown hair. They both had dark silver eyes.

"Who's that over there?" the blue-haired woman asked, pointing to a body in the water.

They moved closer to the dead body, pulled it out and rolled it over.

"Samaari," she said.

"Pathetic. I knew she wouldn't have lasted as queen, Marya," the brunette woman began. "She was never suited to rule from the beginning."

"What about that new girl? She has no authority to take over," Marya said.

"She is of royal blood, though," she replied.

"Does she know who her mother is?"

"No."

"Keep it that way."

Marya bent down to Samaari's corpse and noticed a black twine in her fist. Samaari had clenched her fingers around it. Marya forced Samaari's fingers open, snapping the bones, and pulled out the necklace through the hole in Samaari's palm.

"One necklace, two to go," Marya said, dangling it.

"Give it some time. Jasira won't last either," the brunette woman said.

"And the prophecy? Does she know it was about your son, Prianaj?"

"Yes. But she doesn't know why we need her son alive," Prianaj said, nodding her head and pursing her lips.

"It's a shame you made me wait seven years for this."

"Patience is a virtue Marya. Just not one of yours."

END OF BOOK ONE

BROKEN CARDS OF ACES AND

ALIXEM ALICE

A SHORT STORY

T HE NIGHT CRAWLED ACROSS the marble floor of her home in the wealthy side of The Border. It was as if the darkness found a pristine place to settle in its comfort using the home's breath as its source of life. If it weren't for the thick wool socks Alixem wore to bed every night, her toes would have frozen clean off. The cold was never in her favour. She knew it would never be, even if she wanted it.

Her red hair covered her face when she slept. It was just the way she rested on her silk pillow — face planted in the centre. Alixem's body shot up as she choked on saliva between breaths, and her blood rushed through to her fingertips.

She walked around her mother's house in the dark every night. There was something about the darkness that consumed her thoughts. It was unusual how she found it so peaceful; normally, everyone in The Border ran away from the dark.

"What are you doing up this late?" a voice said from the living room.

Alixem inched closer to the noise, but the room was too dark to see anything. The electricity had been cut from their home for a few days. Electricity was rare; there wasn't enough to go around to everyone in the city.

"Mother?" she asked as she squinted her eyes to make out her silhouette. "I should ask you the same thing."

Her mother, Rayvin, struck a match and lit the candle in front of her. Matches were difficult to obtain in The Border; nobody made them, and they were far too expensive to barter. The candlelight was barely bright enough to light up the room, but the marble furniture reflected it onto her face.

Rayvin was extremely thin and had sharp fingers, almost like claws sticking out from her palms. She always sat on the velvet couch with her left leg crossed over her right. Her arms were folded, and her face remained emotionless. She was thin from exhaustion.

"Pack your belongings, Alixem," she said softly as the air scratched her throat.

"Where are we going?" Alixem hesitated.

Rayvin turned to face her daughter, who had moved to the other side of the room to sit on her armchair and smiled without her eyes. Alixem practically owned that chair and would never let anybody else sit on it.

"The citadel," she said.

"We are perfectly fine here, mother. There's no need to leave."

"The queen needs more Cerberi," Rayvin said as she uncrossed her arms and ran her fingernails through her hair.

The Cerberi were the queen's guards. They patrolled each of the two cities in Diektra — the last surviving nation in the world after the Culling.

"So you are going to the citadel to become a Cerberi. Don't you think you are a little too old for that?" Alixem asked.

"Not me. You," Rayvin responded as she tilted her head down and lifted her eyes.

Alixem's heart sank in her chest, and she pressed her back against the armchair.

"No. That isn't happening," she said firmly, trying to hide the fear in her voice.

Rayvin reached for a deck of playing cards on the floating glass table opposite her. She shuffled them carefully without looking at them. Then she slowly took out all of the king and the jack cards, which should have never been included, and left them on the table. She removed the ace of spades and held onto it with both hands.

"My job is to ensure Queen Samaari gets her Cerberi. The others above me take children from the infirmary immediately after the Procedure and hand them to her as if they are worth nothing. I can't do that."

The Procedure is the modern method of childbirth since the Culling wiped out all the men. It is an arduous method using bone marrow and cells, but the child born can only be female. No male human had been born in two hundred years.

"Just tell her that I don't want to be a Cerberi," Alixem insisted.

"You don't know what she is like. There are things that Queen Samaari has done that are unforgivable."

Alixem covered her ears with her palms flat against her head and brought her knees close to her chest.

"I don't want to hear any more of it," she said.

Rayvin slammed her fists on the table, and a hairline crack ran from the impact site to the other end of the glass. Alixem took her hands away from her head.

"You must, Alixem," she said intensely as the candlelight reflected brightly in the whites of her eyes.

"Mother," Alixem responded calmly, "I am twenty-nine years old. I will make my own decisions."

"The Cerberi have threatened to take away our home. We will lose our status."

Alixem nodded her head slowly in realisation.

"Must be nice knowing status is more important than your daughter," she said as she walked back to her room.

Alixem threw anything she could fit into a throw bag she swore she'd only use in an emergency. She waited for her mother to fall asleep on the couch, and at the sound of her snoring, snuck into the living room.

The playing card in her mother's hand caught Alixem's attention — the ace of spades. She cautiously slid it from her mother's fingertips as not to wake her and stared at it for hardly a second too long before tearing it in half and leaving it on her mother's lap.

Alixem moved to the outskirts of the city, where she found it difficult to adjust. For the most part, she slept at the back of streets with water dripping on her head and rubble occasionally falling around her. It was nowhere near the luxury she was so used to, but twenty-nine years of her mother's despotism took too much of a toll on her.

Most of the buildings were run down, several skyscrapers had collapsed and everything was covered in over-grown vines. Only the freshly refurbished centre walkway through the city was clean. Nobody took care of the outskirts, and it wouldn't make a difference if anybody did.

The condensation Alixem exhaled from her mouth fogged up the scratched acrylic windows she used as a mirror on the days she could make out her reflection. The shattered glass pieces that were scattered across the concrete crunched underneath the Cerberi's boots rushing through.

Alixem poked her head around the side of the derelict house she stayed in temporarily to see all the commotion. At least ten Cerberi congregated around the front of a building she recognised. It was where Vaika used to live. Alixem used to be close with Vaika when she was younger, but over the years, they grew apart. Neither of them was to blame; Rayvin wouldn't allow her daughter to spend her time in the outskirts. It wasn't a good look for her.

Alixem was familiar with the outskirts of The Border, and a lot of the women who lived there recalled seeing her many years ago. The last time she stepped foot near Vaika's home was almost seven years ago, but it has remained empty ever since.

The Cerberi spoke loudly and tried not to reveal too much of what they discovered, but Alixem's curiosity took over. She covered her head with a torn hat that obscured half her face and ran towards the small crowd. She jumped over the tops of the Cerberi's heads, peaking through the gaps and catching a glimpse of a small metal box.

"Nothing to see here," one of the Cerberi said.

"What is that?" Alixem asked.

The Cerberi moved in front of the box to block her view. Alixem smiled mockingly and turned away. She waited behind a broken wall, out of sight but close enough to hear what the Cerberi were saying. She couldn't understand all their words, but she picked up enough information.

"It has the royal engraving," one of the Cerberi said. "But what is it?"

"Who knows?" another one answered. "What do you think we should do with it?"

"The masquerade is next week, perhaps a centrepiece?"

"I'm not sure the queen would want that. Should we tell Orbina about this?"

"We report to her, so we must."

Alixem fled through the centre walkway and underneath The Border's famous underpass to the wealthy side of the city. Tucked between two buildings was a small house. She knocked, and the door swung open.

"Fenix, I'm going to need your help getting ready for the masquerade," Alixem said, not even greeting her friend.

Fenix's face was covered entirely in piercings, and she spoke in broken sentences as her literacy was not well developed.

"You invited. I thought you no more high status," she said.

"My mother wanted me to become a Cerberi, but there was no way I wanted to do that, especially after all the stories you tell me."

"Good choice. My name is on list for masquerade, but I not going."

"Have you heard anything from my mother?" Alixem asked, as she brushed dust from the countertop.

"Alixem, I thought you knew. Rayvin died two days ago from heart attack."

Alixem didn't know whether to laugh or cry — even though she was free from her mother's grasp forever now, she still adored her. Although she felt her throat tighten, her breaths became shallow. Her vision hazed as her eyes tried to force a tear, but she held it back and blinked the blurriness away.

About The Author

"Everybody can write a story. Not everyone can tell your story the same way you do." — G. A. Jøhn.

When he's not writing young-adult fiction and fantasy, G. A. Jøhn watches movies, plays video games, and reads all the books he can. He lives in Sydney, Australia and is currently working on more exciting stories to share.

Visit him at www.gajohnbooks.com and on Instagram at @gajohn_author for all news and updates.

CPSIA information can be obtained
at www.ICGtesting.com
Printed in the USA
LVHW092020020122
707637LV00011B/99/J